BASIC/NOT BORING MATH SKILLS

GEOMETRY & MEASUREMENT

Grades 4–5

Inventive Exercises to Sharpen Skills and Raise Achievement

Series Concept & Development
by Imogene Forte & Marjorie Frank

Exercises by Sheri Preskenis

Illustrations by Kathleen Bullock

Incentive Publications, Inc.
Nashville, Tennessee

About the cover:

Bound resist, or tie dye, is the most ancient known method of fabric surface design. The brilliance of the basic tie dye design on this cover reflects the possibilities that emerge from the mastery of basic skills.

Cover art by Mary Patricia Deprez, dba Tye Dye Mary®
Cover design by Marta Drayton, Joe Shibley, and W. Paul Nance
Edited by Jennifer E. Janke

ISBN 0-86530-433-5

1 2 3 4 5 6 7 8 9 10 07 06 05 04

PRINTED IN THE UNITED STATES OF AMERICA
www.incentivepublications.com

TABLE OF CONTENTS

CELEBRATE BASIC MATH SKILLS

Basic does not mean boring! There is certainly nothing dull about . . .

. . . sharpening geometry skills by solving mysteries in the locker room
. . . using measurement skills to get rock climbers up a wall, make posters for a pep rally, or help the coaches straighten out some mixed-up boxes of uniforms
. . . finding geometric figures at the ballpark or on the bottom of the swimming pool
. . . searching for angles at a gymnastics meet and a diving competition
. . . figuring out which fans at the ball game have eaten the most snacks
. . . going to the wrestling mat to sharpen your skills with circles
. . . tracking down a lost basketball that fell off the bus on the way to a game

These are just a few of the interesting adventures students can explore as they celebrate basic math skills with geometry and measurement. The idea of celebrating the basics is just what it sounds like—sharpening math skills while enjoying the fun of sports teams and events. Each page of this book invites students to practice a high-interest math exercise built around sports situations at a middle school with great school spirit. This is not just any ordinary fill-in-the-blanks way to learn. These exercises are fun and surprising, and they make good use of thinking skills. Students will do the useful work of practicing a specific geometry or measurement skill, while stepping into the fascinating world of team sports. They will tackle clever problems with athletes, coaches, fans, and equipment.

The pages in this book can be used in many ways:

* to sharpen or review a skill with one student
* to reinforce the skill with a small or large group
* by students working on their own
* by students working under the direction of a parent or teacher

Each page may be used to introduce a new skill, reinforce a skill, or assess a student's ability to perform a skill. And there's more than just the great student activities. You will also find an appendix of resources helpful to students and teachers—including a ready-to-use test for assessing geometry and measurement skills.

As your students take on the challenges of these adventures with geometry and measurement, they will grow! And as you watch them check off the basic math skills they have strengthened, you can celebrate with them!

The Skills Test

Use the skills test beginning on page 56 as a pretest and/or a post-test. This test will help you check the students' mastery of problem-solving skills and strategies, and will prepare them for success on achievement tests.

SKILLS CHECKLIST FOR
GEOMETRY & MEASUREMENT, Grades 4-5

✔	SKILL	PAGE(S)
	Identify kinds of lines: perpendicular, parallel, and intersecting lines	10, 11
	Identify and describe points, lines, line segments, rays, and planes	10, 11, 12, 13
	Identify different kinds of angles	10, 12, 14, 15
	Identify plane figures	13, 17, 18
	Identify congruent angles	16
	Identify and define kinds of triangles: (scalene, equilateral, isosceles, right)	17, 18
	Identify and define different kinds of polygons	19, 20
	Identify define, and distinguish among different kinds of quadrilaterals	21
	Identify properties and parts of a circle	22
	Identify similar and congruent figures	23
	Identify symmetrical figures and lines of symmetry	24
	Identify transformations in plane figures	25
	Identify characteristics or space figures	26
	Recognize and define space figures	26, 27
	Compare volume of space figures	28
	Identify various U.S. customary units for measuring	29
	Compare and convert among U.S. customary measurements	30
	Use U.S. customary measurements for measurement tasks	31
	Identify various metric units for measuring	32
	Use metric units for measuring length	33
	Choose the correct tool for a measurement task	34
	Find the circumference of circles	35
	Find the perimeter of plane figures	36, 37
	Find the area of circles	38, 40
	Find the area of plane figures	39, 40
	Use metric units to find the volume of cubes and rectangular prisms	41
	Use U.S. customary units to find the volume of cubes and rectangular prisms	42, 43
	Use formulas to find perimeter, area, and volume	44
	Use measurement skills and scale to determine distances on a map	45
	Estimate the measurements of angles	46
	Determine the measurements of angles	47
	Determine time measurements	48, 49, 50

GEOMETRY & MEASUREMENT

Skills Exercises

Identify Points, Lines, Angles, Rays, & Planes

LOCKER ROOM MYSTERY

The lockers in the Ashland Middle School locker room have unusual names on them! Can you figure out which locker belongs to which athlete?

Write the letter of the locker that matches each clue.

A ANGLE
B LINE
C PARALLEL LINES
D RAY
E PERPENDICULAR LINES
F LINE SEGMENT
G INTERSECTING LINES

Clues

D	1. Ashley's Locker	I'm part of a line that has only one endpoint.
C	2. Sara's Locker	We are lines in the same plane, but we never met.
A	3. Gina's Locker	I am made of two rays that have the same end point.
F	4. Kate's Locker	I am part of a line that has two endpoints.
B	5. Gayle's Locker	I extend in opposite directions without end.
E	6. Kayla's Locker	When I meet another line, it is always at a right angle.
G	7. Megan's Locker	The two of us meet and cross each other.

8. Draw a pair of perpendicular lines.	9. Draw a pair of intersecting lines.	10. Draw a pair of parallel lines.

Name

SIGNS FROM THE CROWD

The rowdy crowd is getting ready for the opening football game of the season between the Ashland Grizzlies and the Crescent City Cougars.

Draw a line from each label to match the correct sign.

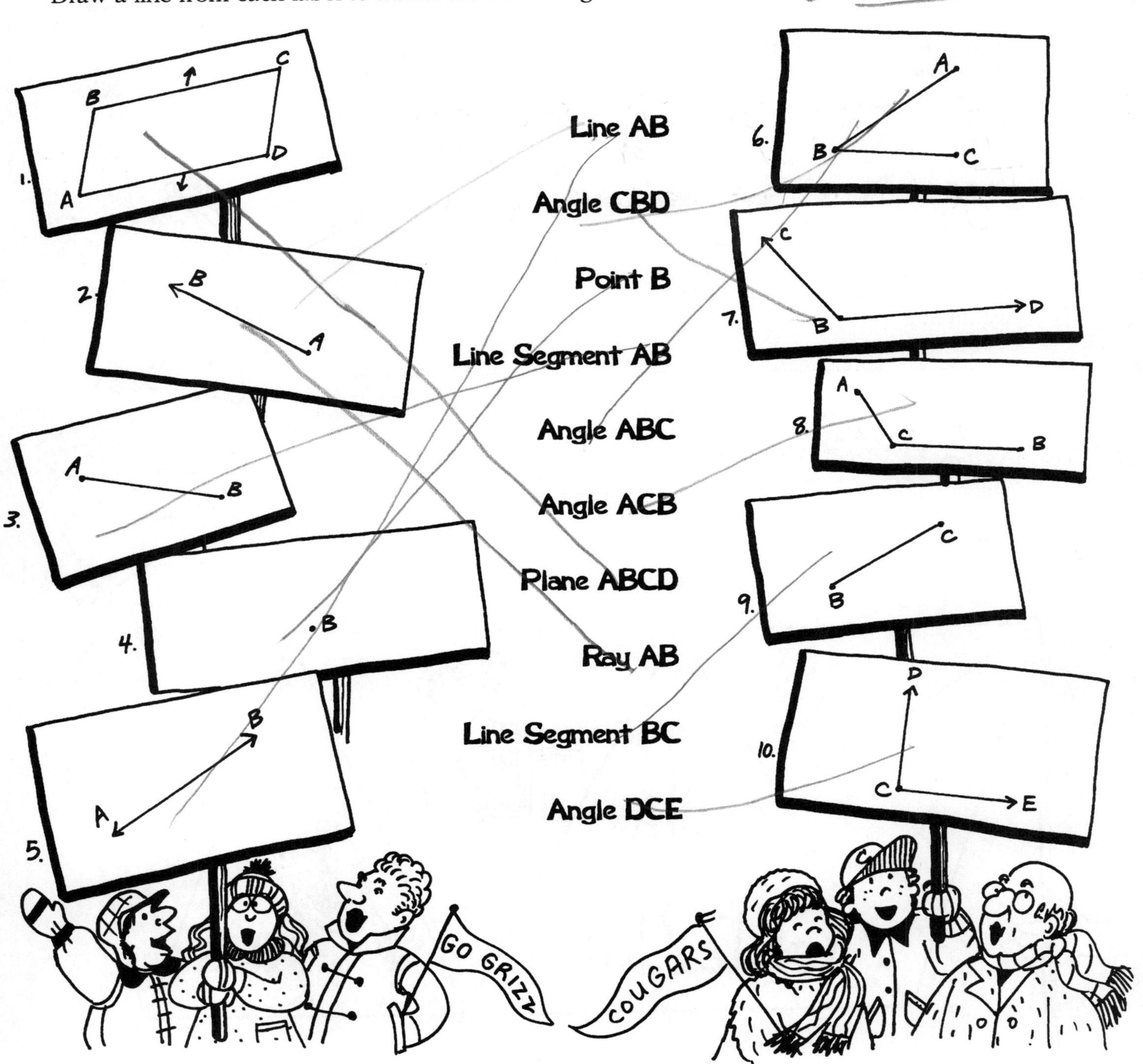

Name

BANNER GEOMETRY

Geometry shows up in all kinds of places, even the new Grizzly team banner offers a good place for a geometry review! Color the banner with the Grizzly colors of red, white, and light blue. (You decide where the colors look best.) Then answer the geometry questions.

1. How many line segments can you find? ______
   List the line segments here.

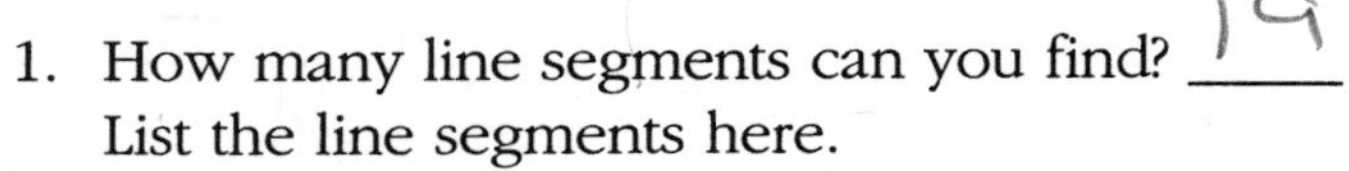

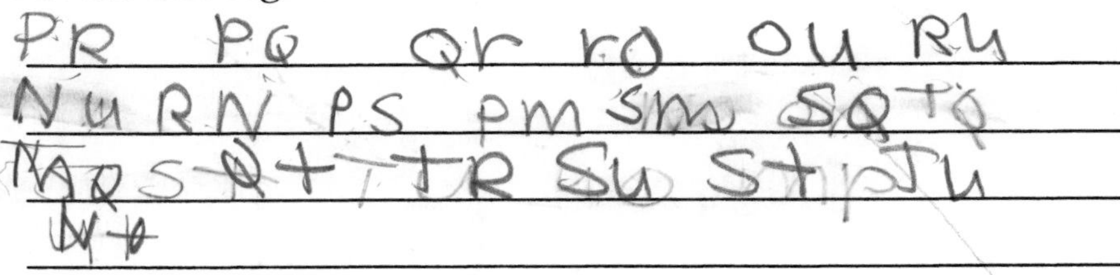

2. How many angles can you find? ______
   List them here.

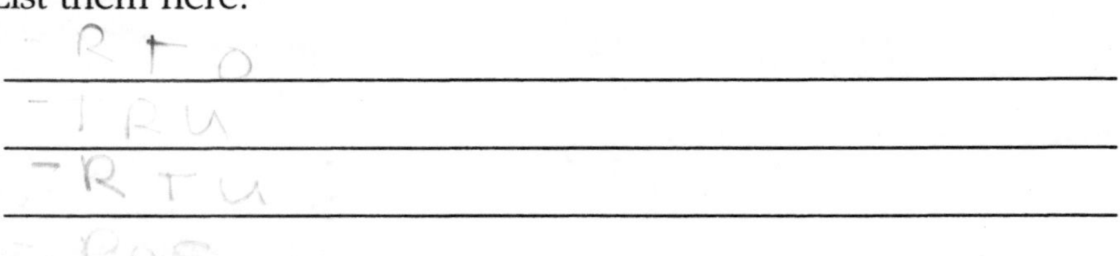

3. Write your initials in the vertex of angle RNO.

Name ________________________________

GEOMETRY AT THE BALL PARK

Take your markers to the ballpark for the first home game, and search the scene for geometric places and spaces.

Color or trace at least three of each figure.
Use the color chart to find the right color for each one.

FIGURE	COLOR
point	**blue**
plane	**green**
line segment	**red**
angle	**yellow**

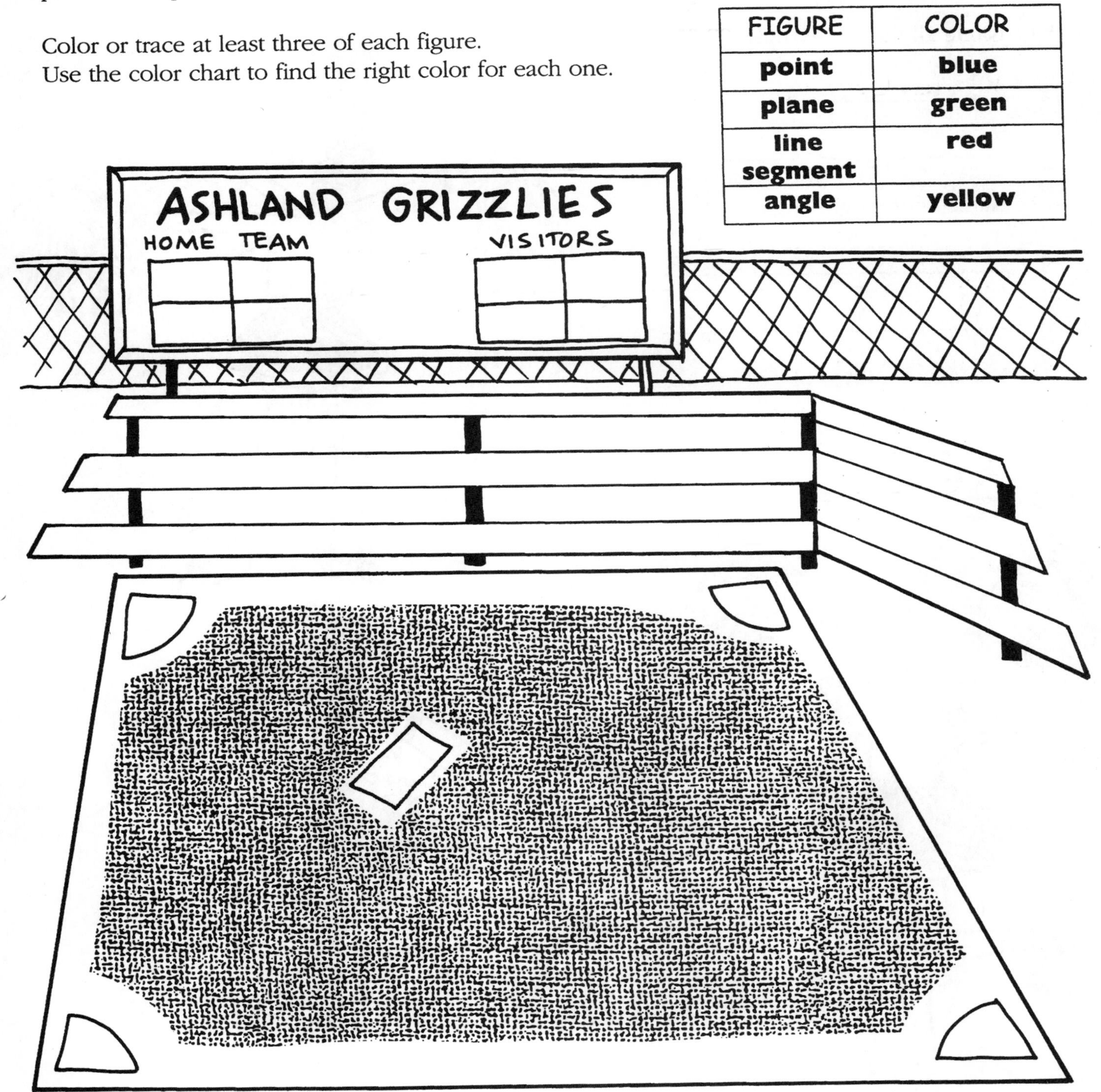

Name

Identify Kinds of Angles

NEW ANGLE ON CHEERS

The cheerleaders are practicing for the first pep rally. They're practicing some new tricks and routines.

1. Name all the acute angles.

   F K O

2. Name all the right angles.

   B H J L

3. Name at least two obtuse angles.

   A G O P E C

Name

WHAT'S THE ANGLE?

It looks as if some of the cross country ski team members did a poor job of putting away their skis. The skis are criss-crossing each other at many different angles.

Get out your crayons or markers and show that you can identify the kinds of angles.

Trace at least 6 acute angles in RED.

Trace at least 10 right angles in BLUE.

Trace at least 5 obtuse angles in GREEN.

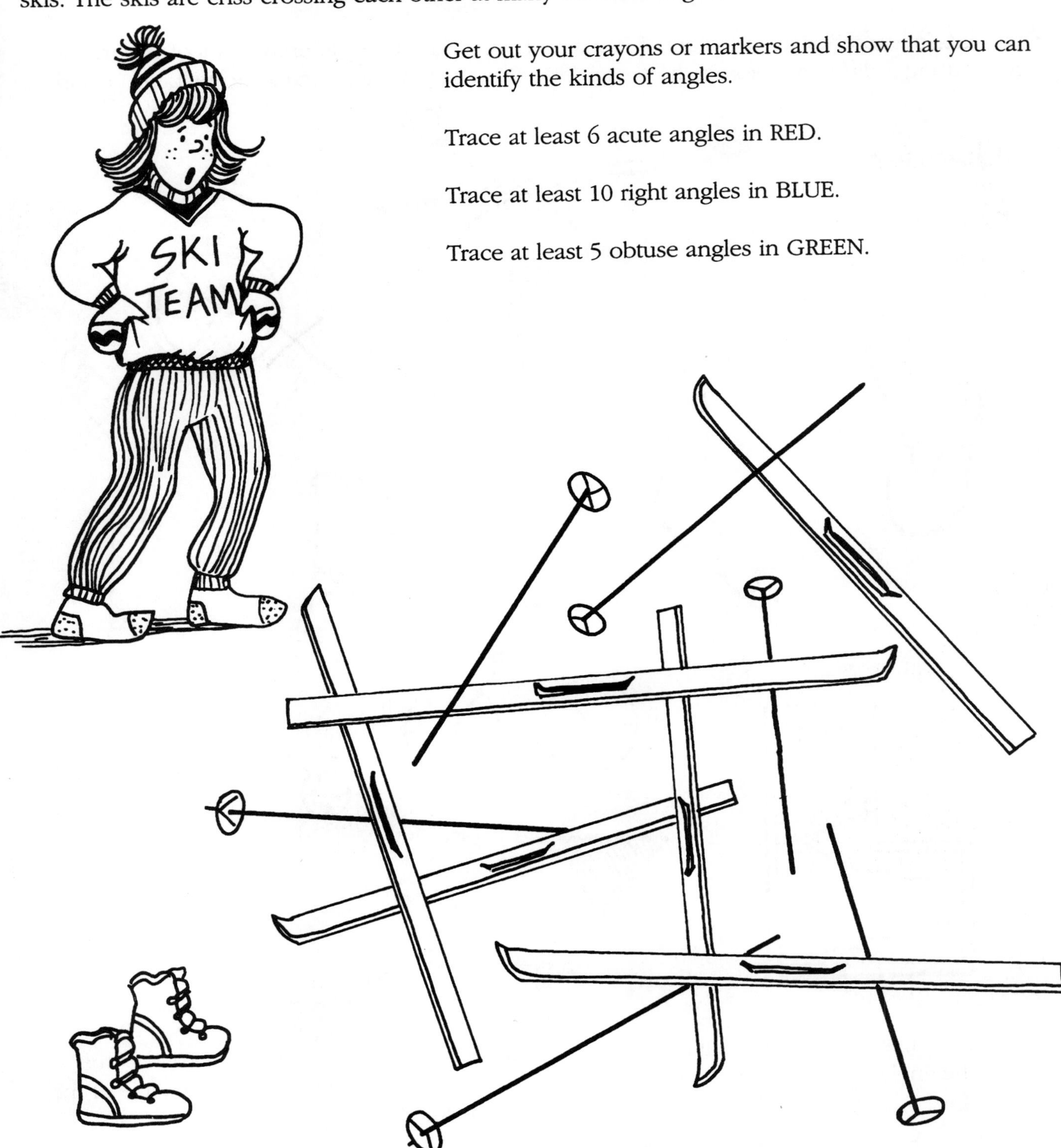

Name

Identify Congruent Angles

ANGLES IN EQUIPMENT

Angles show up in lots of different sports equipment. Take a look at these items, and find pairs of congruent angles.

Choose 6 (or more) different colors of markers. Trace 6 (or more) different pairs of congruent angles, using a different color to show each pair. (You may use a protractor to help you decide.)

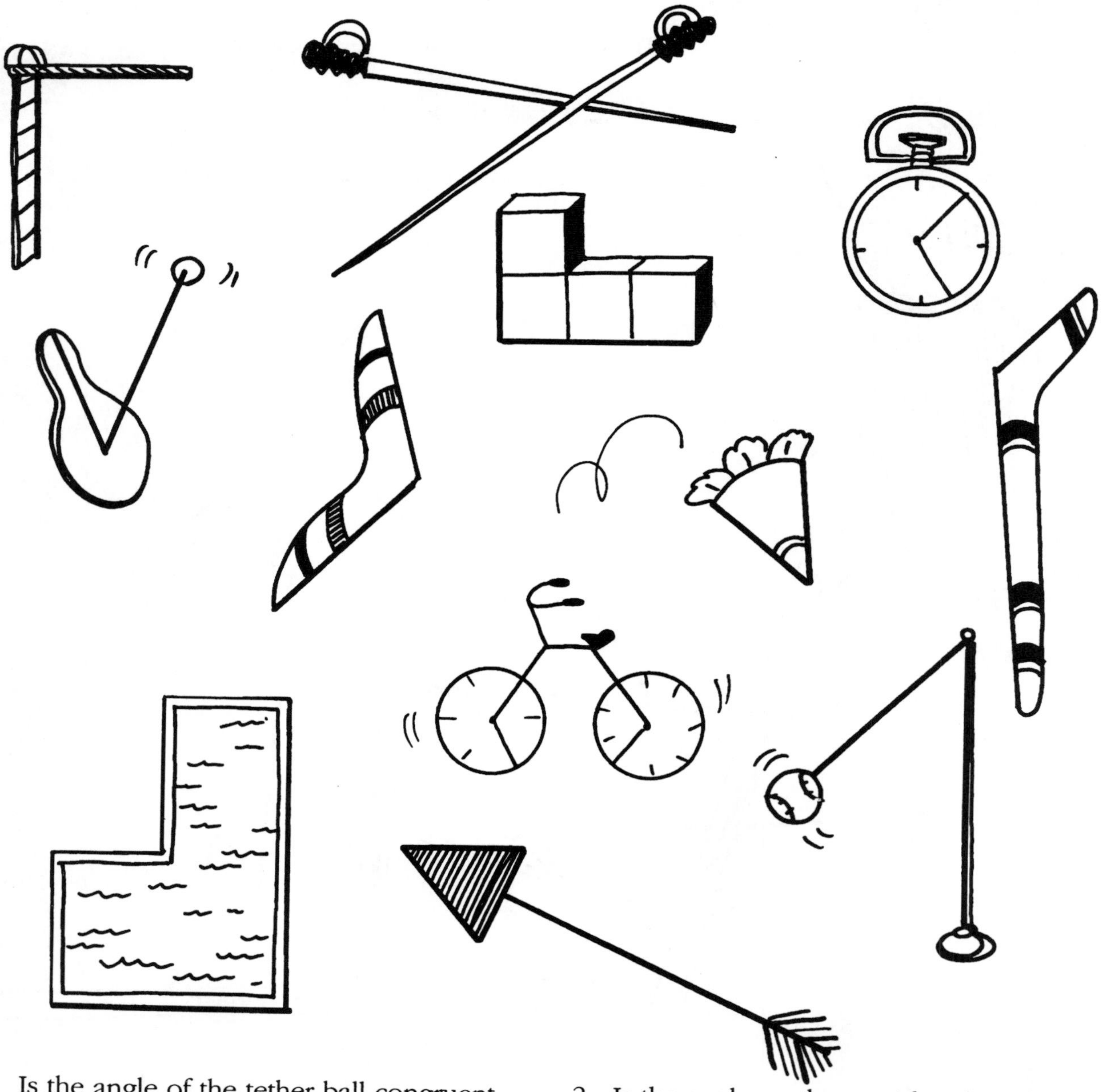

1. Is the angle of the tether ball congruent to the angle at the tip of the arrow?

____

2. Is the angle on the outside edge of the boomerang congruent to the angle of the paddle ball?

____

Name

Kind of Triangle	*Color*
right triangles	blue
isosceles triangles	green
scalene triangles	purple
equilateral triangles	red
other shapes	yellow

Identify Kinds of Triangles

GYM FLOOR GEOMETRY

The Booster Club spent a year collecting money for a new gym floor. What a job! Student groups were asked to submit designs for the floor. Here is the prizewinning design. It's full of triangles. Can you find them all?
Follow the chart to color the floor design.

Name

Identify Plane Figures

A PLANE MESS

Coach Jackson teaches math when he is not coaching volleyball. He had some great posters ready for his geometry lesson today, but, as usual, he forgot to close the window. A huge wind blew his stuff all over the floor.

Get the definition posters back together with the math terms in time for class. Draw a line from each math term to its matching poster.

Oh, my!

isosceles triangle

A. An 8-sided polygon

B. A 6-sided polygon

polygon

hexagon

rectangle

C. A 5-sided polygon

D. A quadrilateral with only 1 pair of parallel sides

trapezoid

scalene triangle

E. A rectangle with 4 equal sides

F. A parallelogram with 4 right angles

octagon

G. A quadrilateral whose opposite sides are parallel

rhombus

obtuse triangle

equilateral triangle

H. A polygon with 4 sides

I. A triangle with 1 obtuse angle

triangle

J. A triangle with all sides equal

K. A triangle with 1 right angle

square

pentagon

L. A triangle with no equal sides

right triangle

N. A plane figure made up of sides joined at endpoints

M. A plane figure with 3 sides

O. A triangle with 2 equal sides

quadrilateral

parallelogram

P. A parallelogram with 4 equal sides

Name

Identify Kinds of Polygons

SHARP EYES FOR SHAPES

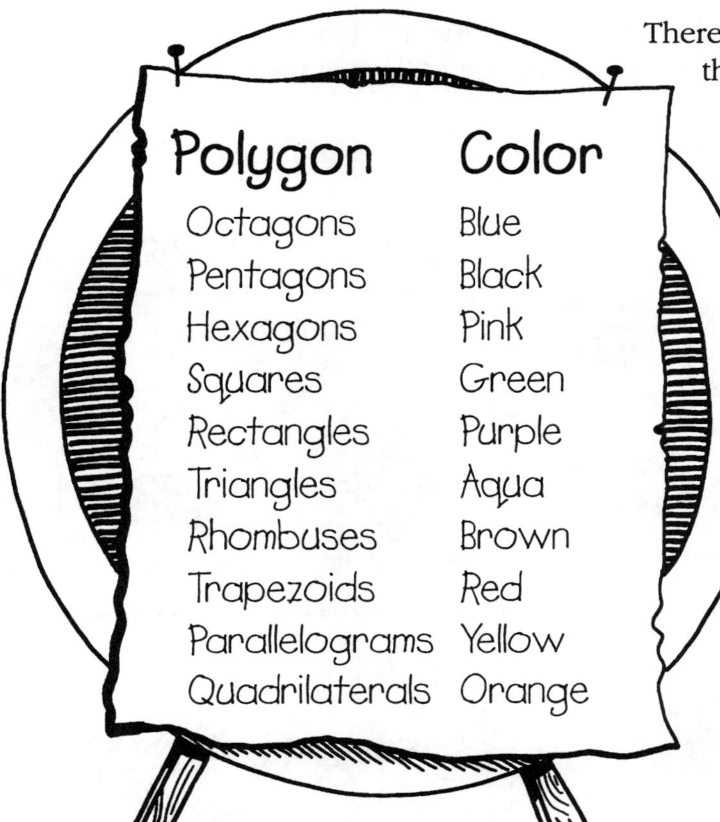

Polygon Color
Octagons Blue
Pentagons Black
Hexagons Pink
Squares Green
Rectangles Purple
Triangles Aqua
Rhombuses Brown
Trapezoids Red
Parallelograms Yellow
Quadrilaterals Orange

There is a piece of hidden sports equipment in this picture. Do you know what it is?

To find out, follow the color chart to color each figure. Because some shapes fit more than one definition, color each shape in the order indicated on the chart.

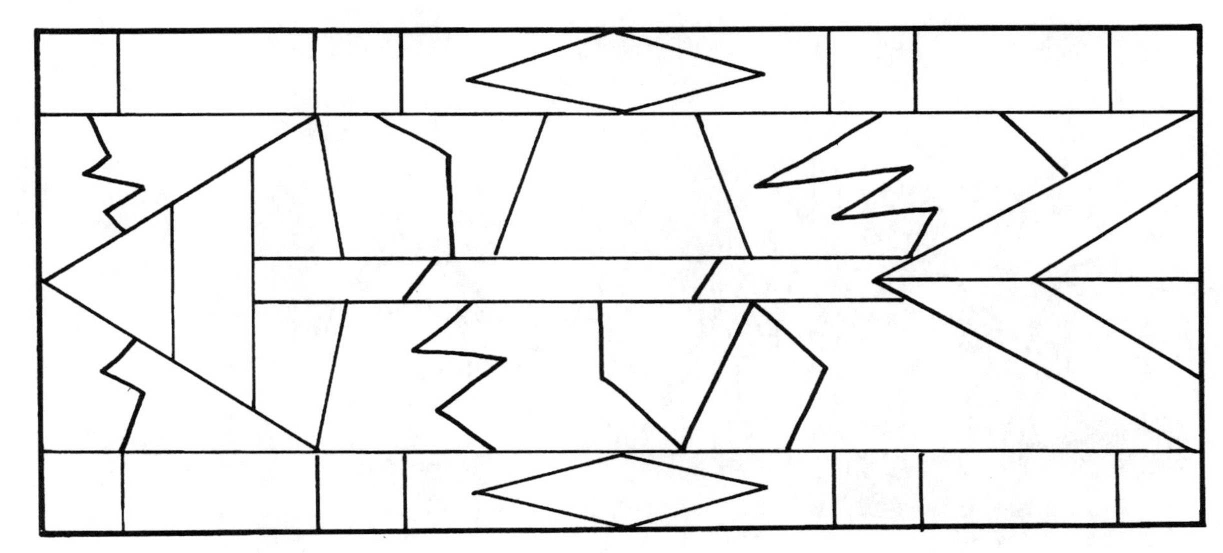

1. What is it? ____________________
2. In what sport is this object used? ____________________

Name ____________________

Identify Kinds of Polygons

THE GREAT SHAPE MATCH-UP

All the parents of the basketball players have come to watch the first home game. They're all holding up numbered cards with symbols to match the names of their kids on the team.

Search the cards to match the parents and players. Write the number of each card on the correct player.

Name

KEEPING BUSY

Ashley never stops being active in sports! As soon as one season is over, she starts something new. So she has many labels: athlete, basketball player, volleyball player, tennis player, gymnast, pitcher, and swimmer.

Quadrilaterals are like that, too. They have many labels.

All quadrilaterals have four sides. But a four-sided figure can show up in many "uniforms" or different "looks."

Which figures match each description? (There may be more than one.)

trapezoid
square
quadrilateral
parallelogram
rhombus
rectangle

1. All angles are right angles, but all sides are not equal. ________________
2. Only one pair of opposite sides is parallel. ________________
3. A rectangle with all sides equal ________________
4. A figure with two pairs of opposite sides parallel ________________
5. A parallelogram with all sides the same length ________________

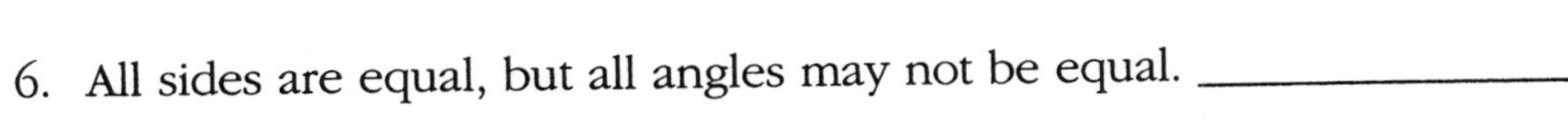

6. All sides are equal, but all angles may not be equal. ________________

Write (T) TRUE or (F) FALSE next to each statement.

_____ 1. All squares are rectangles.
_____ 2. All rectangles are quadrilaterals.
_____ 3. No rhombuses are trapezoids.
_____ 4. Rectangles have no right angles.
_____ 5. A rectangle is a square.
_____ 6. A square is a rectangle.
_____ 7. All squares are rhombuses.
_____ 8. A trapezoid is a quadrilateral.
_____ 9. All parallelograms are rectangles.
_____ 10. All rhombuses are squares.

Name

Identify Properties and Parts of Circles

GOING IN CIRCLES

The wrestling team is warming up for their big match against their rivals, the Crescent City Cougars. The athletes will show off their wrestling skills on a circular mat with a diameter of 11 meters.

Use the circle diagram of the mat to show off your geometry skills and knowledge about the parts of a circle.

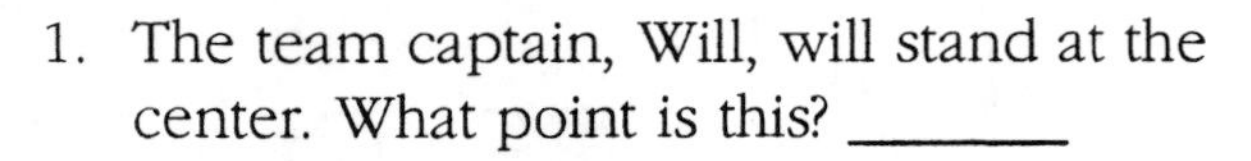

1. The team captain, Will, will stand at the center. What point is this? _______

2. Jason and Dan warm up by jogging back and forth on the diameters. Name the diameters.

   ____

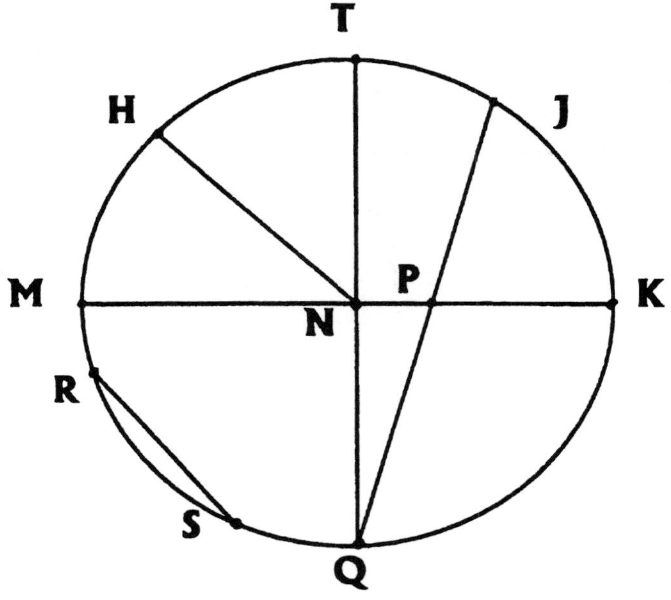

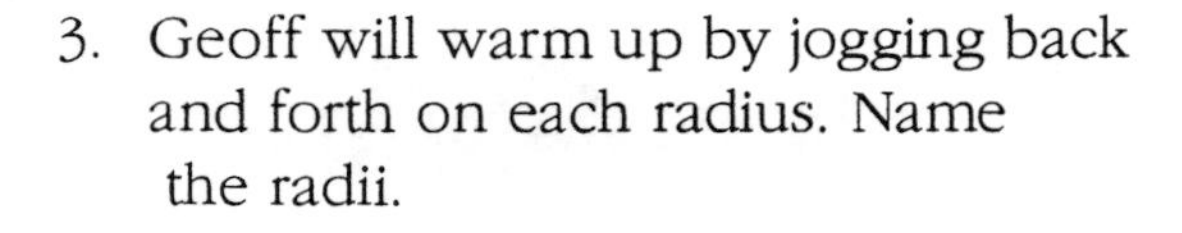

3. Geoff will warm up by jogging back and forth on each radius. Name the radii.

   ____

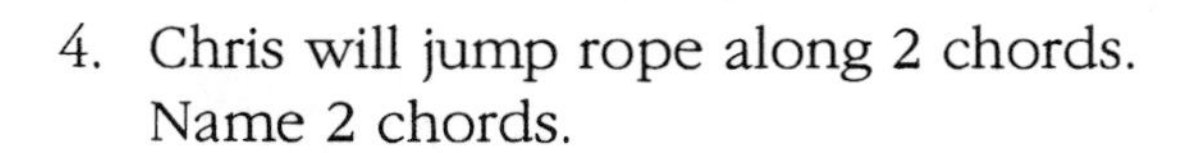

4. Chris will jump rope along 2 chords. Name 2 chords.

   ____

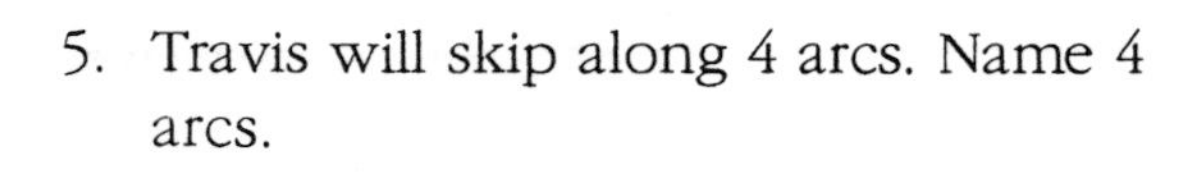

5. Travis will skip along 4 arcs. Name 4 arcs.

   ____

Draw another wrestling mat. Make sure it contains all parts of the circle listed below. Trace them with the colors shown.

1 center (black)

2 radii (red)

2 chords (green)

2 diameters (orange)

2 arcs (blue)

Name

A CLOSE LOOK AT FIGURES

Wrestlers must get weighed before each match. In wrestling, opponents are matched as closely as possible in size and weight. It doesn't matter if they look similar or have similar figures—it's the exact weight that counts!

Geometric figures must be exactly the same size and shape to be called congruent. If they are the same shape, they are similar, no matter what their size.

How would you describe the figures of these two wrestlers? (Circle one answer.)

congruent   similar   neither

Label each pair of figures C (congruent) or S (similar).

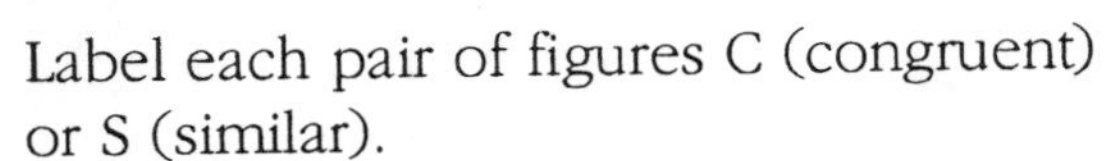

1. 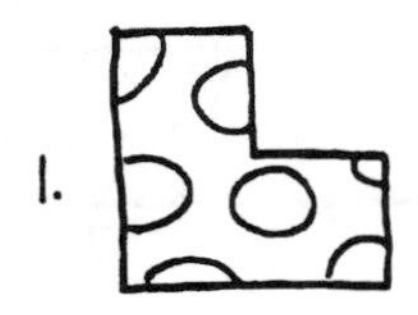 

____

2. 

____

3. 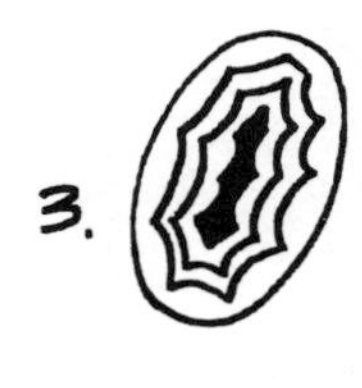 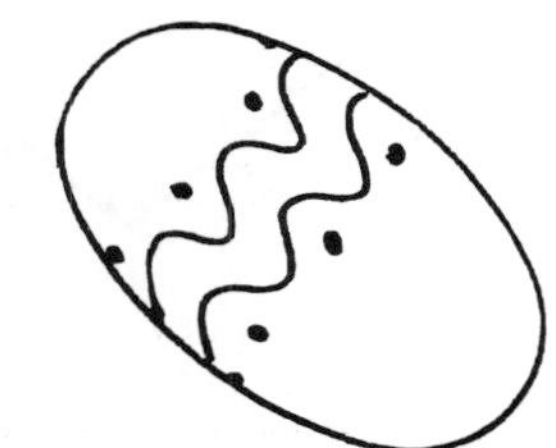

____

4. 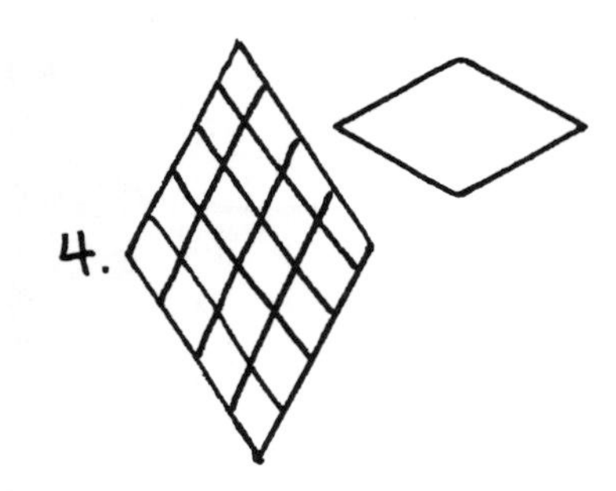

____

5. 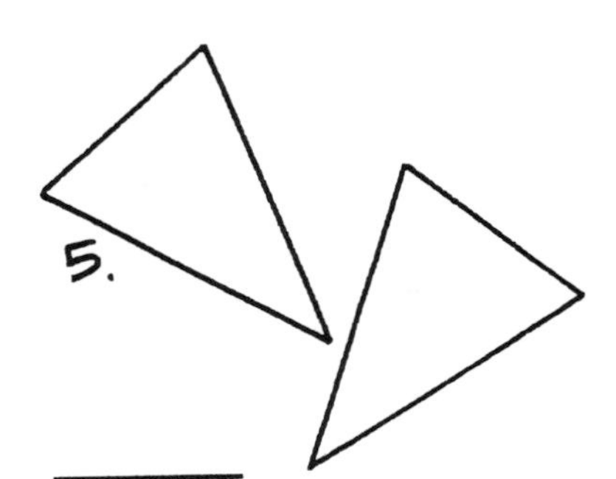

____

6. 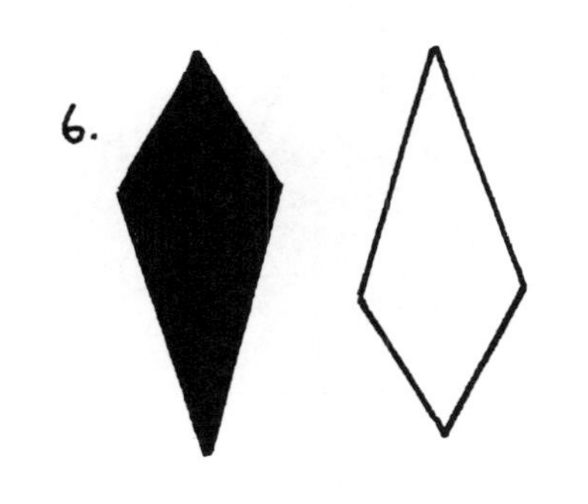

____

7. 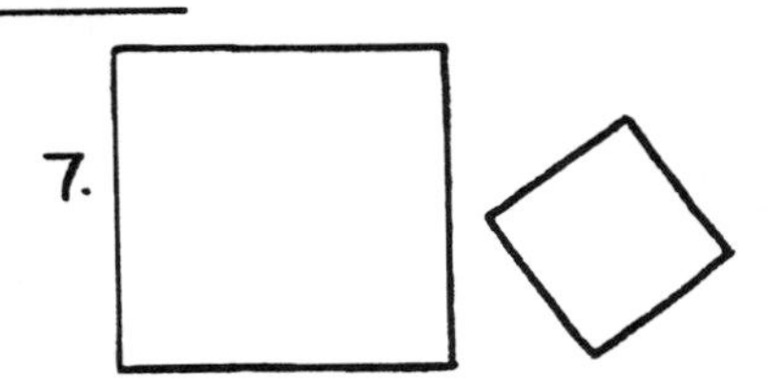

____

8. 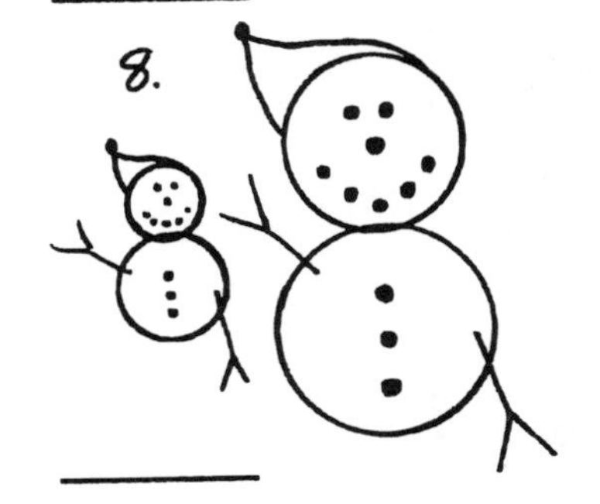

____

9. 

____

Name ____________________

Identify Symmetrical Figures

MIRROR IMAGES

Jenna often practices her dance moves in front of a mirror. She hopes the reflection shows a perfect performance.

In a symmetrical figure, each half is a perfect reflection of the other.

Look at the figures below. Color the ones that are symmetrical. Use a ruler to draw the line of symmetry in each symmetrical figure.

Complete figures I, J, and K to make them symmetrical. The line of symmetry is already given for you.

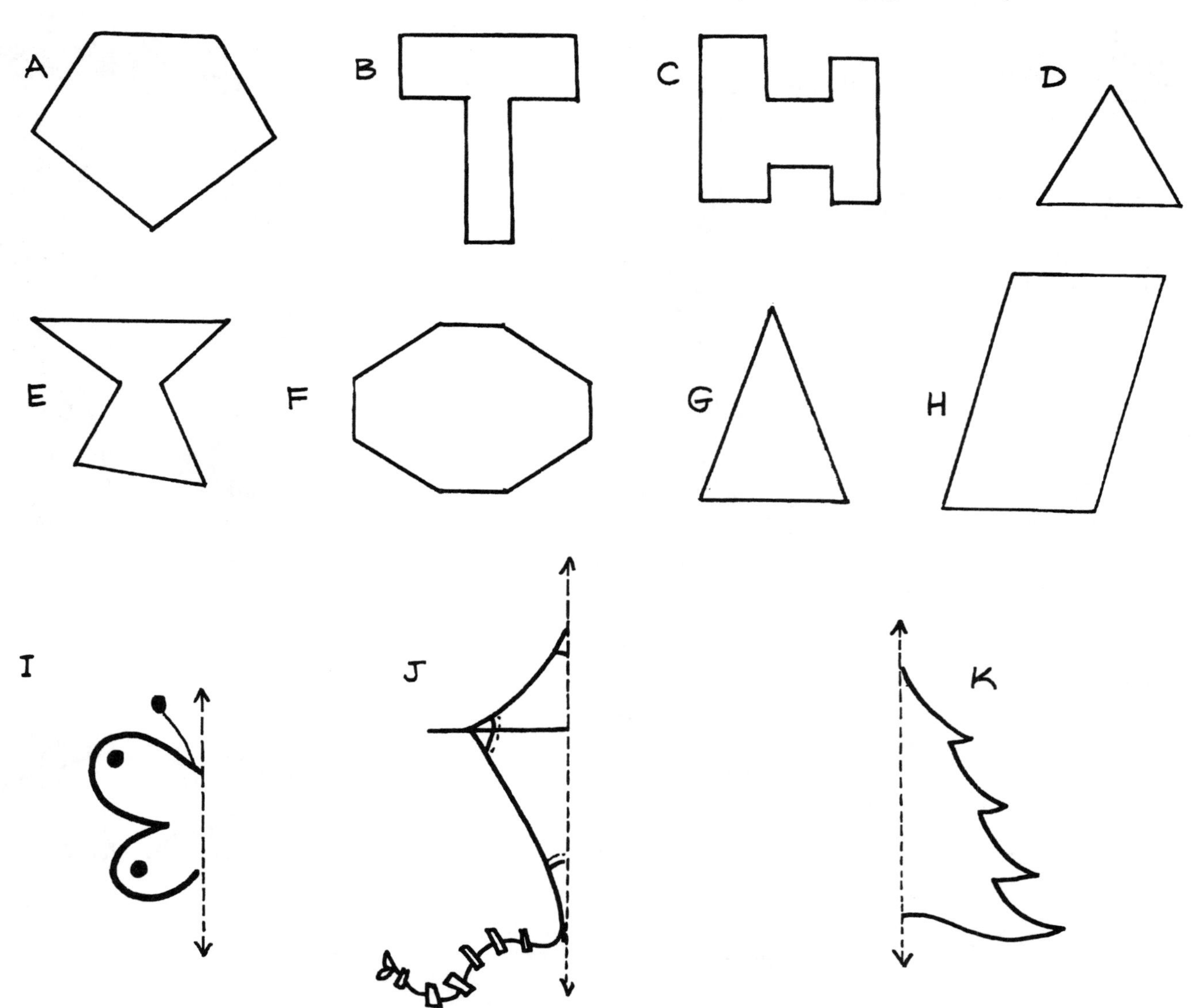

Name

Recognize Transformations of Plane Figures

OUT OF ORDER

The sports storage room is a mess. Pieces of equipment have been tossed around carelessly. Custodian George has to put things back in order.

Look at each pair of items. Tell whether the second item in each pair is a slide, flip, or turn of the first item. Write S, F, or T beside each pair. (Some pairs may have more than one label!)

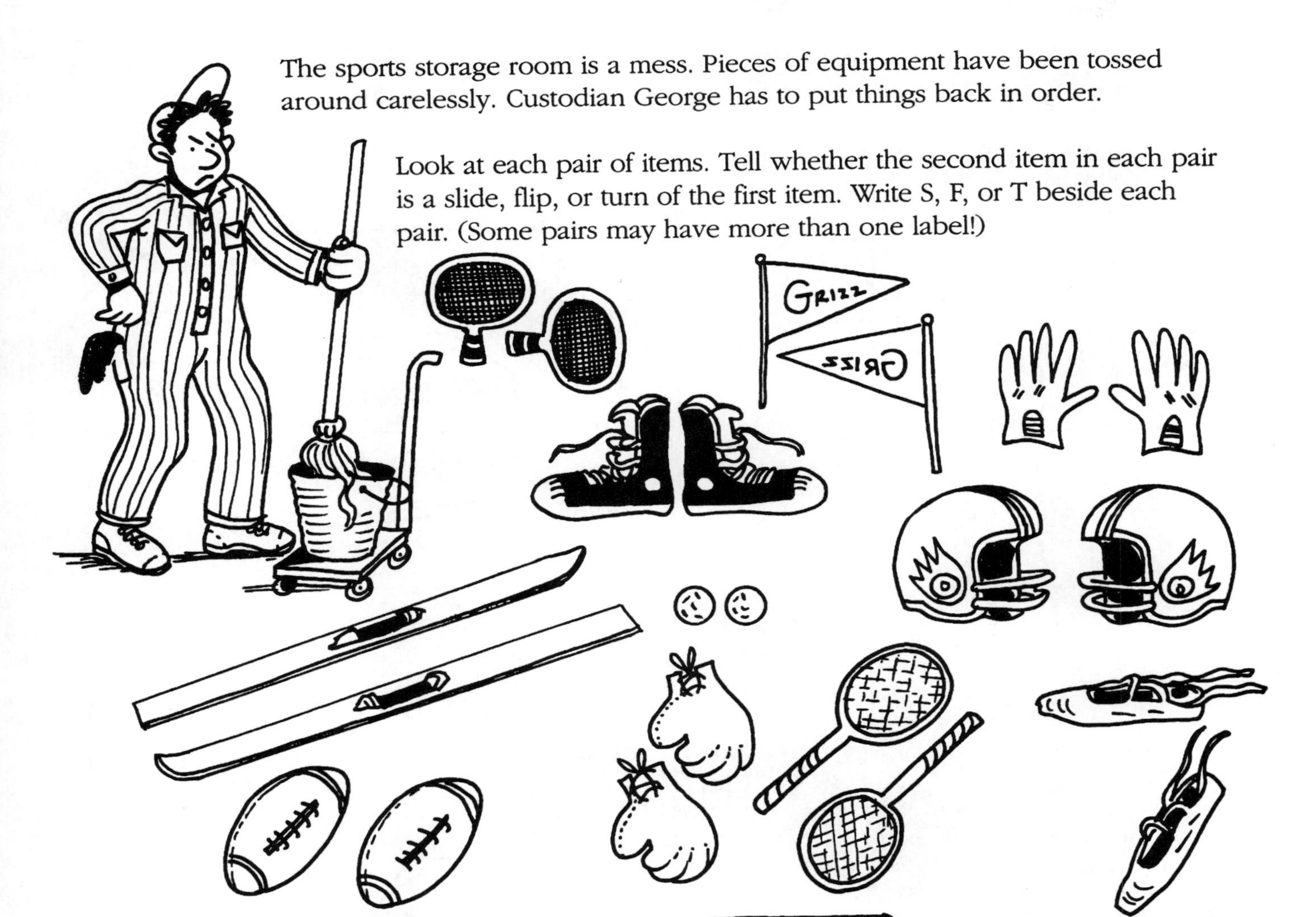

Draw a slide (translation) of this figure:

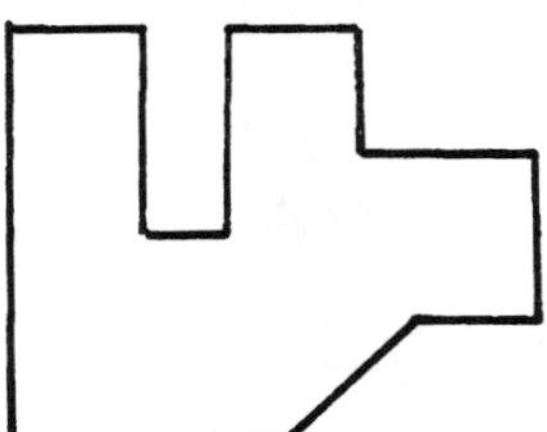

Draw a turn (rotation) of this figure:

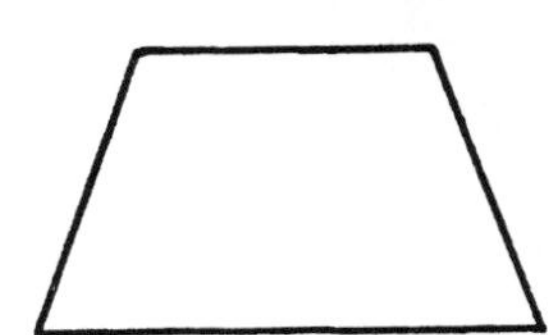

Draw a flip (reflection) of this figure:

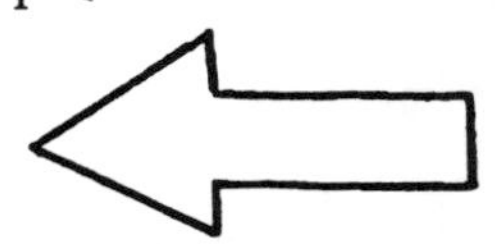

Draw a turn (rotation).of this figure:

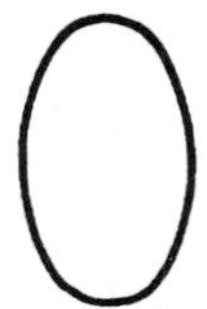

Name

Identify Space Figures

GEOMETRY ON WHEELS

FIGURE	COLOR
cubes	**red**
spheres	**blue**
cones	**green**
rectangular prisms	**yellow**
cylinders	**purple**
triangular prisms	**brown**
pyramids	**orange**

The practice course for the skating team is loaded with geometric space figures. Skaters practice jumps and turns over and around the figures.

Color or outline the figures to identify them. Use the color chart.

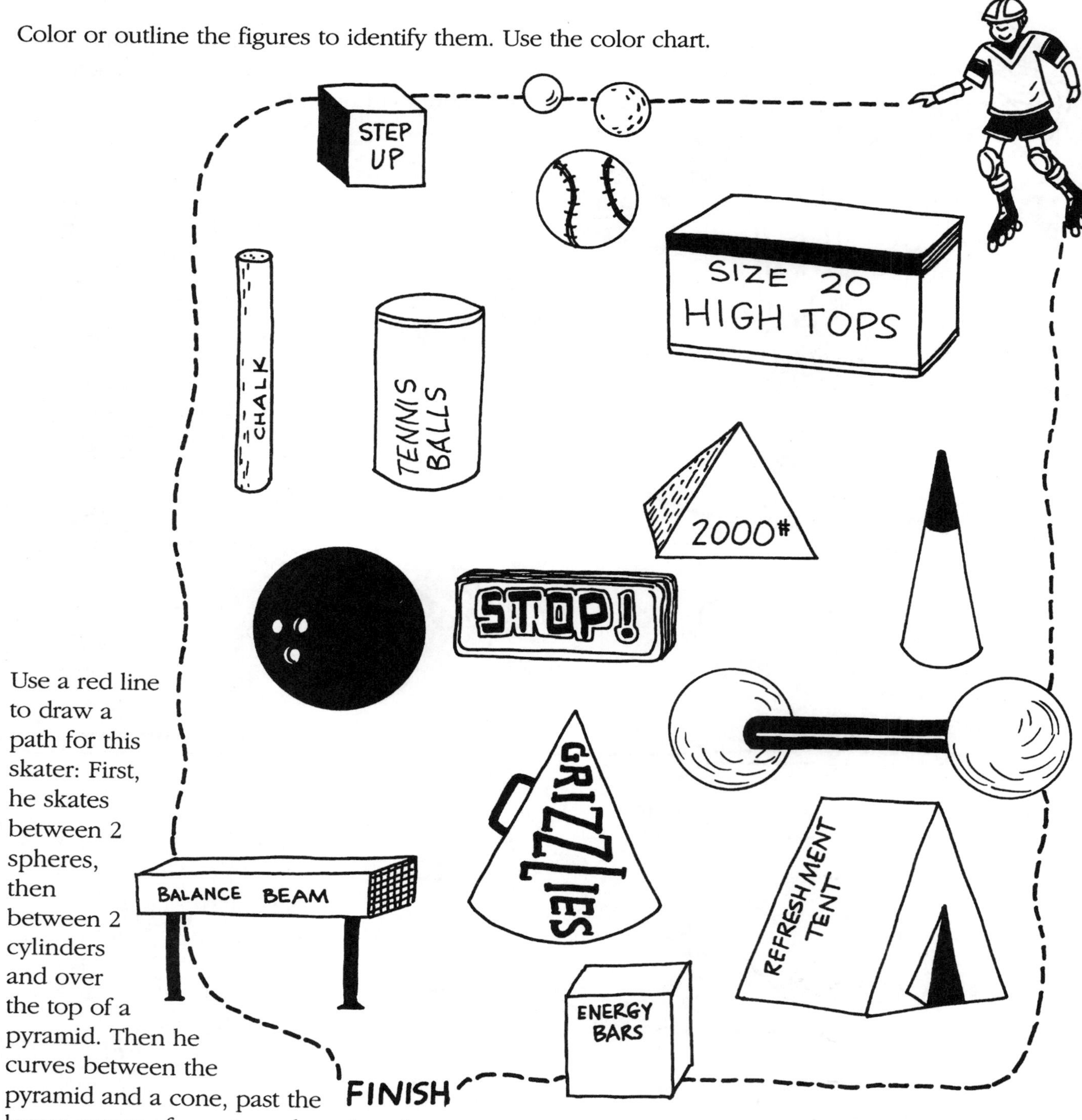

Use a red line to draw a path for this skater: First, he skates between 2 spheres, then between 2 cylinders and over the top of a pyramid. Then he curves between the pyramid and a cone, past the lower corner of a rectangular prism, between a cone and a sphere, and over the top of a triangular prism. He circles around the outside of the prism and jumps over a cube to the finish line.

Name

CLUES ON THE CLIPBOARD

The football coach's clipboard is the source of those very challenging (and secret) plays for the big football game.

These clipboards hold clues to the names of some geometric space figures. Read the clue on each clipboard, then choose a figure from the answer clipboard, and write its name with each clue. Be careful! These are just as challenging as the coach's new football plays.

Name

GET A JUMP ON VOLUME

The skateboard club has some very talented competitors. They practice year-round on ramps and jumps in the new skateboard park. Here are some of the structures they have built to jump over.

Examine each structure to find its volume. (Count the cubic units.) Then answer the questions.

1. Which has 9 cubic units? ____________
2. Which jump's volume is 8 cubic units? ____________

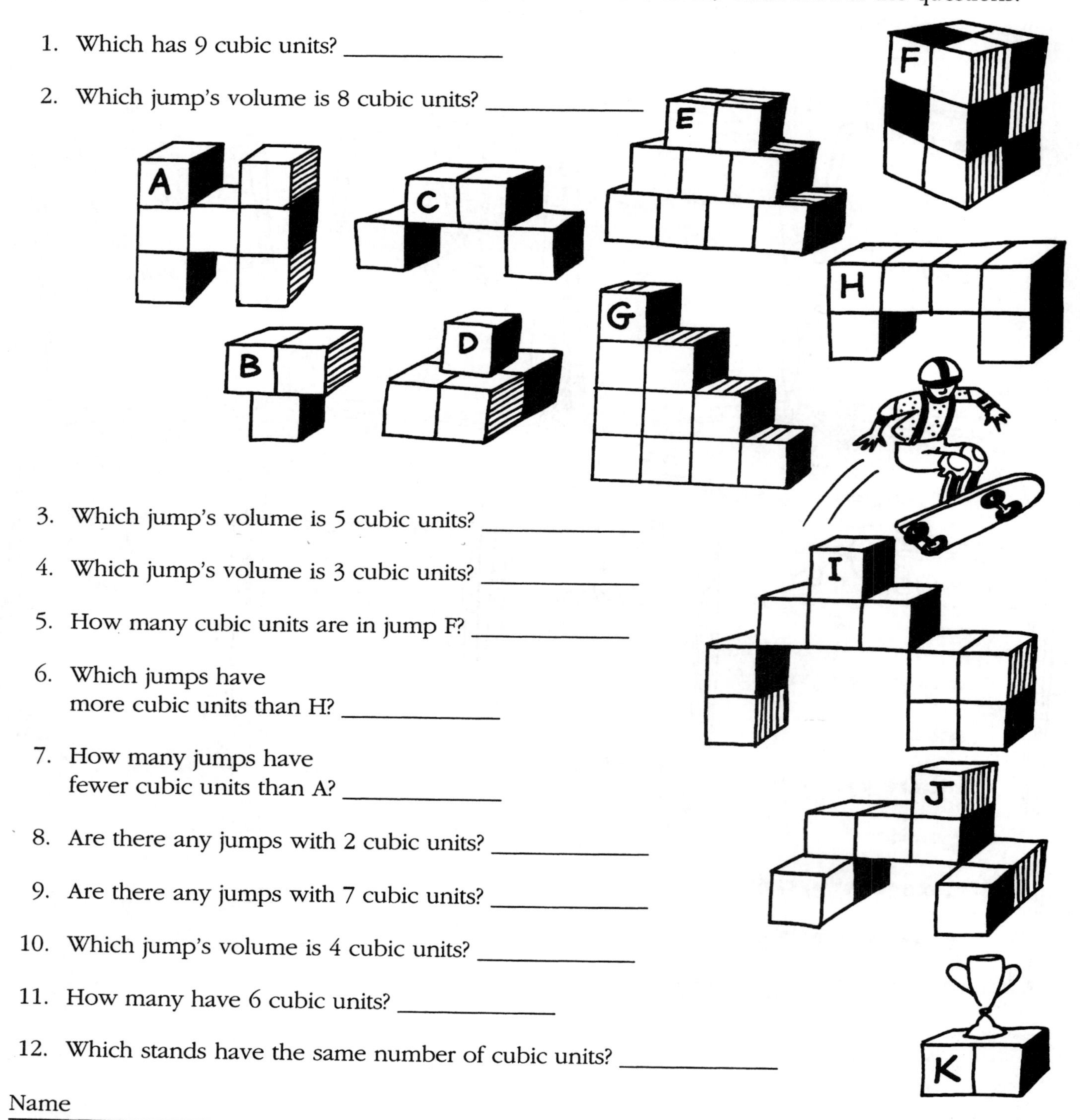

3. Which jump's volume is 5 cubic units? ____________
4. Which jump's volume is 3 cubic units? ____________
5. How many cubic units are in jump F? ____________
6. Which jumps have more cubic units than H? ____________
7. How many jumps have fewer cubic units than A? ____________
8. Are there any jumps with 2 cubic units? ____________
9. Are there any jumps with 7 cubic units? ____________
10. Which jump's volume is 4 cubic units? ____________
11. How many have 6 cubic units? ____________
12. Which stands have the same number of cubic units? ____________

Name

COURTSIDE MEASUREMENTS

Search the tennis court puzzle for U.S. Customary measurement units.

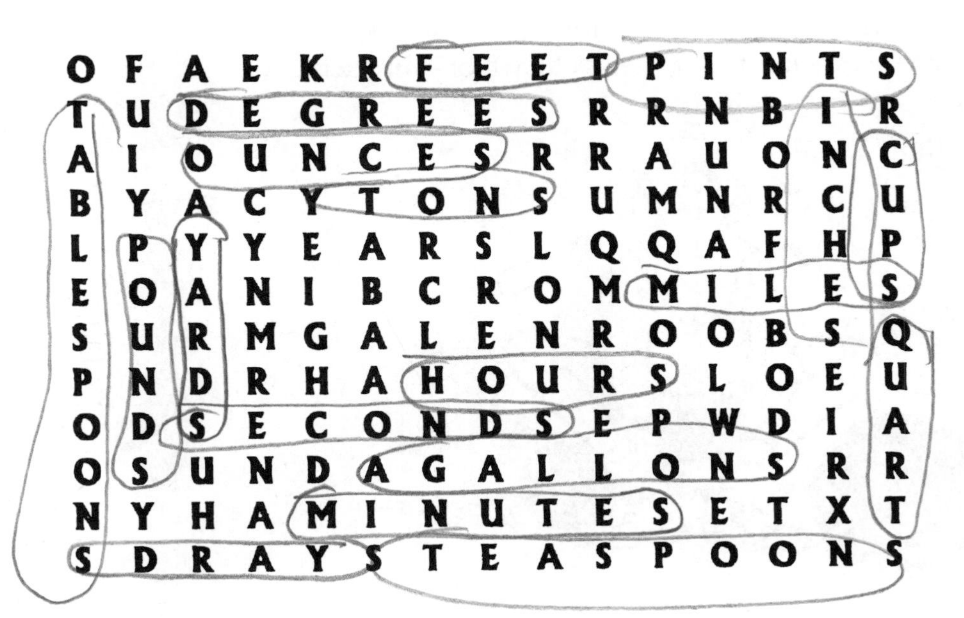

Circle a term in the puzzle to fill each of these blanks. (One is written backwards.)

1. The tennis players are 13 ______ old.
2. Sheri jogs 2 ______ to the tennis court.
3. An hour equals 60 ______.
4. A glass of Thirst Blaster holds 14 ______.
5. Jordan is exactly 5 ______ tall.
6. Four cups of milk is one ______.
7. Danielle weighs 83 ______.
8. Suzie practices for 3 ______ after school.
9. 1 cup of juice holds 16 ______
10. Six feet equals 2 ______.
11. The temperature is 77 ______ today.
12. The match lasted 30 minutes, 10 ______
13. One quart of drink equals 2 ______.
14. One quart of drink equals 4 ______.
15. Jeri's height is 5 feet, or 60 ______.
16. The tennis bus weighs one ______.
17. One tablespoon equals 3 ______.
18. The team drank 40 quarts— that's ten ______.

Name ______

MORE OR LESS?

Ooops! The weight lifters have a problem at practice today. The weights on both ends of the bars are not equal for every athlete!

Look at the measurement amounts below. They have the same problem. Compare the measures. Write > (greater than), < (less than), or = in each circle.

1.	6 pt	○	3 qt
2.	16 oz	○	1 lb
3.	9 qt	○	16 pt
4.	5 pt	○	10 c
5.	3 ft	○	36 in
6.	3 lbs	○	59 oz
7.	2 c	○	1 pt
8.	2 lbs	○	22 oz
9.	7 yds	○	21 ft
10.	2000 lbs	○	1 T
11.	12 ft	○	3 yds
12.	6 gal	○	18 qt
13.	38 in	○	1 yds
14.	4 yds	○	100 in
15.	2 qt	○	8 gal
16.	11 in	○	1 ft
17.	4 T	○	6 tsp
18.	100 sec	○	2 mins
19.	7 hrs	○	420 mins
20.	10 gal	○	40 qt

Name

Use U.S. Customary Units

UNIFORM MEASUREMENTS

It's time to order uniforms for the soccer team. Matt is getting measured to find out exactly what size he needs.

Look at the measurements needed for Matt. Find a measuring tape or ruler, and a friend or classmate. Use U.S. Customary units to find these measurements on your friend. Round all the measurements to the nearest whole unit.

circumference of head = ________

circumference of neck = ________

shoulder to tip of fingers = ______

distance around waist = ________

length of longest finger = ________

distance around thigh = ________

width of kneecap = __________

knee to ankle = ______________

total leg length = ____________

distance around ankle = ________

foot length heel to toe = ________

Name ______________________________

Identify Metric Units of Measure

PASSING THE TEST

In most schools, athletes must keep good grades in order to play a school sport. How is Tom doing on his measurement test? He needs to have 9 correct in order to pass the test.

Circle the numbers of the correct answers. Cross out the wrong answers, and replace them with the correct answers.

Measurement Test

Student Name: *Tom* Date: *January 6*

1. Liquids are measured in *meters*.
2. Would 5 millimeters of water fill a cup? yes (no)
3. Circle the greater amount: (10 kilograms) 100 grams
4. Could someone's hand be 1 decimeter long? (yes) no
5. Circle the larger amount: 2 kilometers (200 meters)
6. 1 meter = *100* centimeters.
7. 1 kilometer = *1000* meters.
8. *100* milliliters = 1 liter.
9. 100 meters = *1* decimeter(s).
10. 5 grams = *5000* milligrams.
11. 20 meters = *2* centimeters.
12. *300* centimeters = 3 meters.
13. 1 metric ton = *1000* kilograms.
14. 1 gram = *100* milligrams.
15. 10 kilometers = *500* meters.
16. 10,000 milligrams = *10* grams.

Will Tom pass? ________

Name

CLIMBING THE WALL

The Adventure Club uses a climbing wall at the gym to practice rock climbing skills. Draw a path up the wall for each of these climbers. The path must connect the hand-hold points.

After you draw the paths, use a centimeter ruler to measure the paths. Then find the length of the path to the nearest meter, using this scale: 1 centimeter of measure = 1 meter on the wall.

Length of Michael's path ________

Length of Marcy's path ________

Name ______________________________

MEASUREMENTS ON PARADE

The Booster Club members are building floats and getting ready for a big sports parade. Help them decide which tools they should use for different measurement tasks. Choose from among the tools listed on the float below. There may be more than one correct answer.

What tool should be used to measure . . .

1. the amount of paint needed for floats? ____________________
2. the length of the float? ____________________
3. the angle of the streamers? ____________________
4. the temperature of the air on parade day? ____________________
5. the height of each float? ____________________
6. the weight of the popcorn? ____________________
7. the time the parade will take? ____________________
8. the amount of glue needed for decorations? ____________________
9. the weight of the band's instruments? ____________________
10. the width of the band leader's hat? ____________________
11. the amount of soda pop needed for the workers? ____________________
12. the length of the drummer's drumsticks? ____________________
13. the height of the tuba in the band? ____________________
14. the time it takes for the band to play the opening song? ____________________
15. the weight of the team members riding the float? ____________________

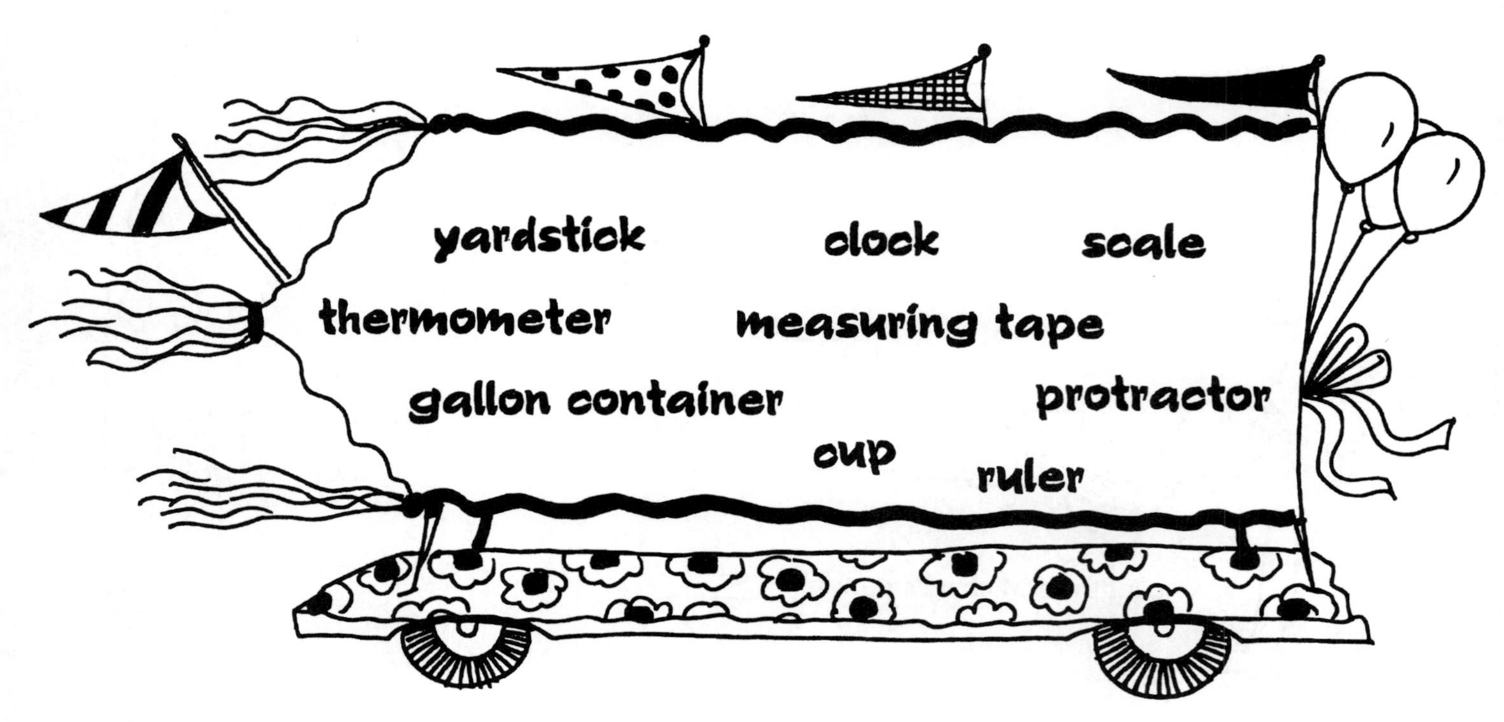

Name ______________________________

CIRCLES EVERYWHERE YOU LOOK

Circles show up all the time in the world of sports. See below the circles of different sizes athletes find in their sports.

Use the formula for circumference ($C = \pi d$) for each circle.

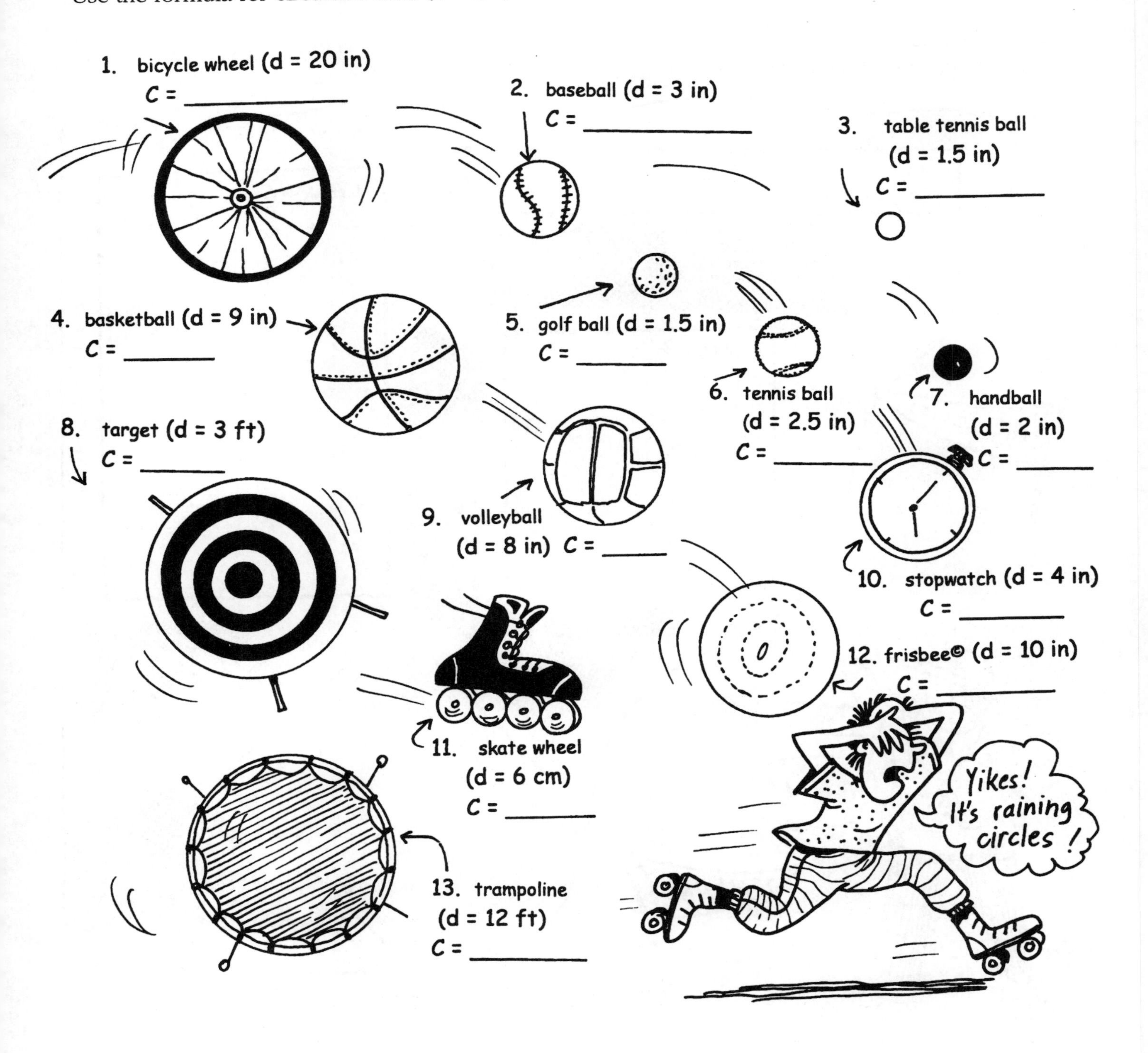

Name

Find Perimeter of Space Figures

AROUND THE EDGE

The sports at Ashland Middle School take place in all kinds of places and spaces. But it seems that no matter where the practices are located, one thing is always the same. At every practice, every coach asks the athletes to warm up by running around the outside of the field, room, mat, or court!

Figure out how far the athletes have to run at each of these locations. Find the perimeter of each sports area shown. Write the perimeter inside the area. Write P = ________.

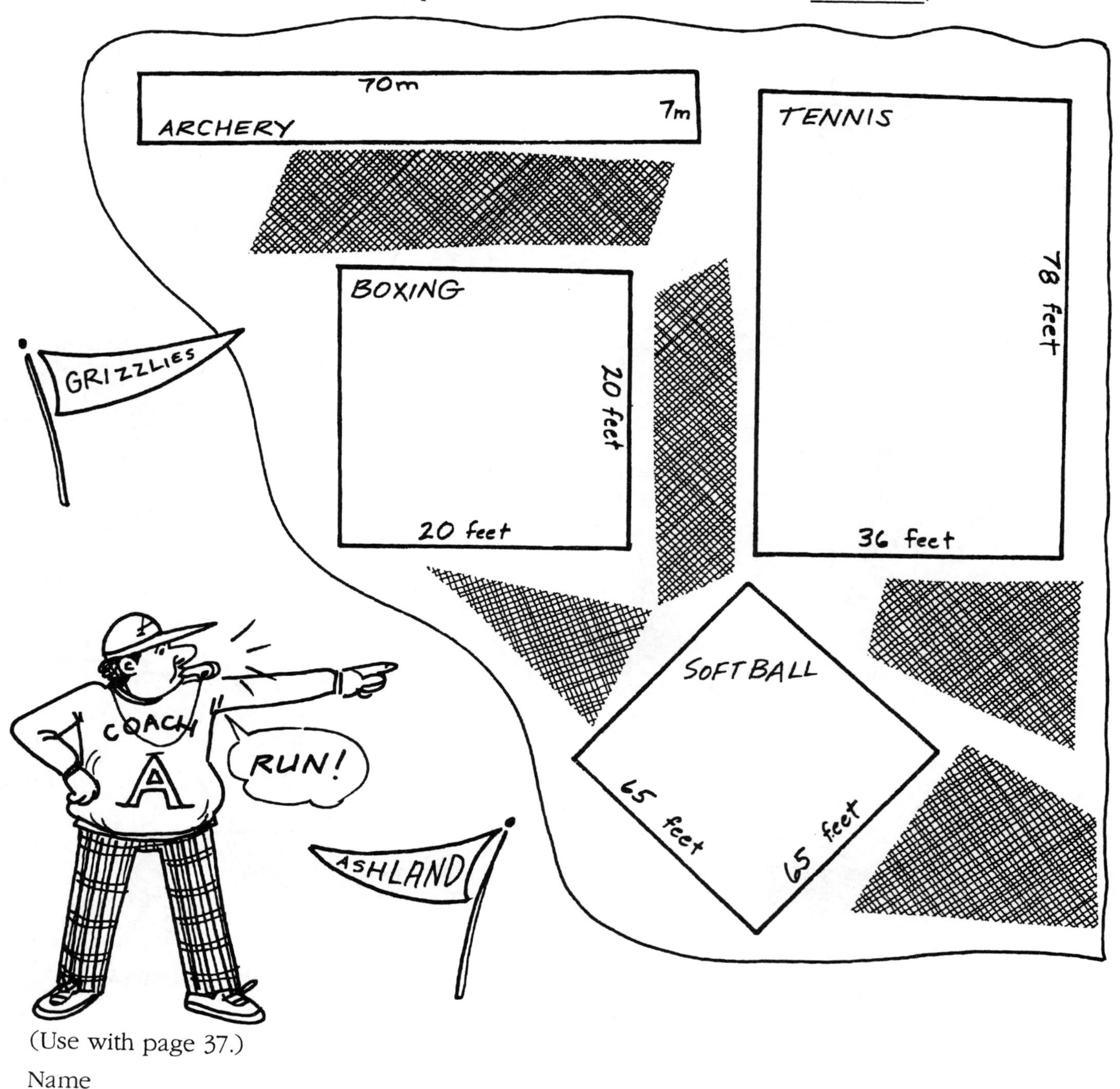

(Use with page 37.)

Name

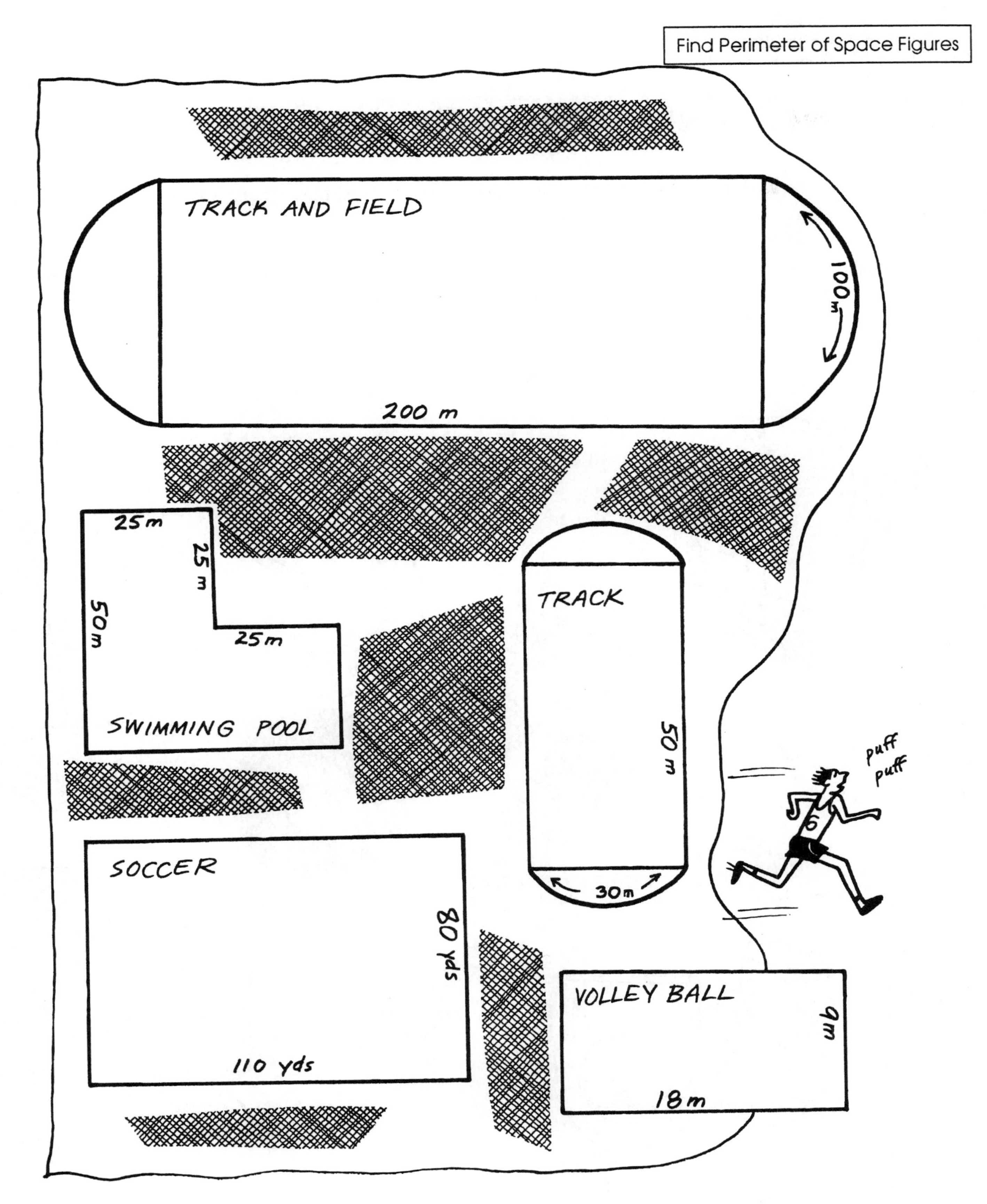

In which sport do the athletes run the farthest to warm up? ______________________

(Use with page 36.)

Name ____

Find Area of Circles

WATCHING THE TIME

Time plays a part in most sports. Coaches, fans, referees, and players are always watching the clock.

Take a good look at these clocks. Then find the area of the face on each clock.
Write the answer (A = __________) near each clock. Make sure you label your answers correctly.

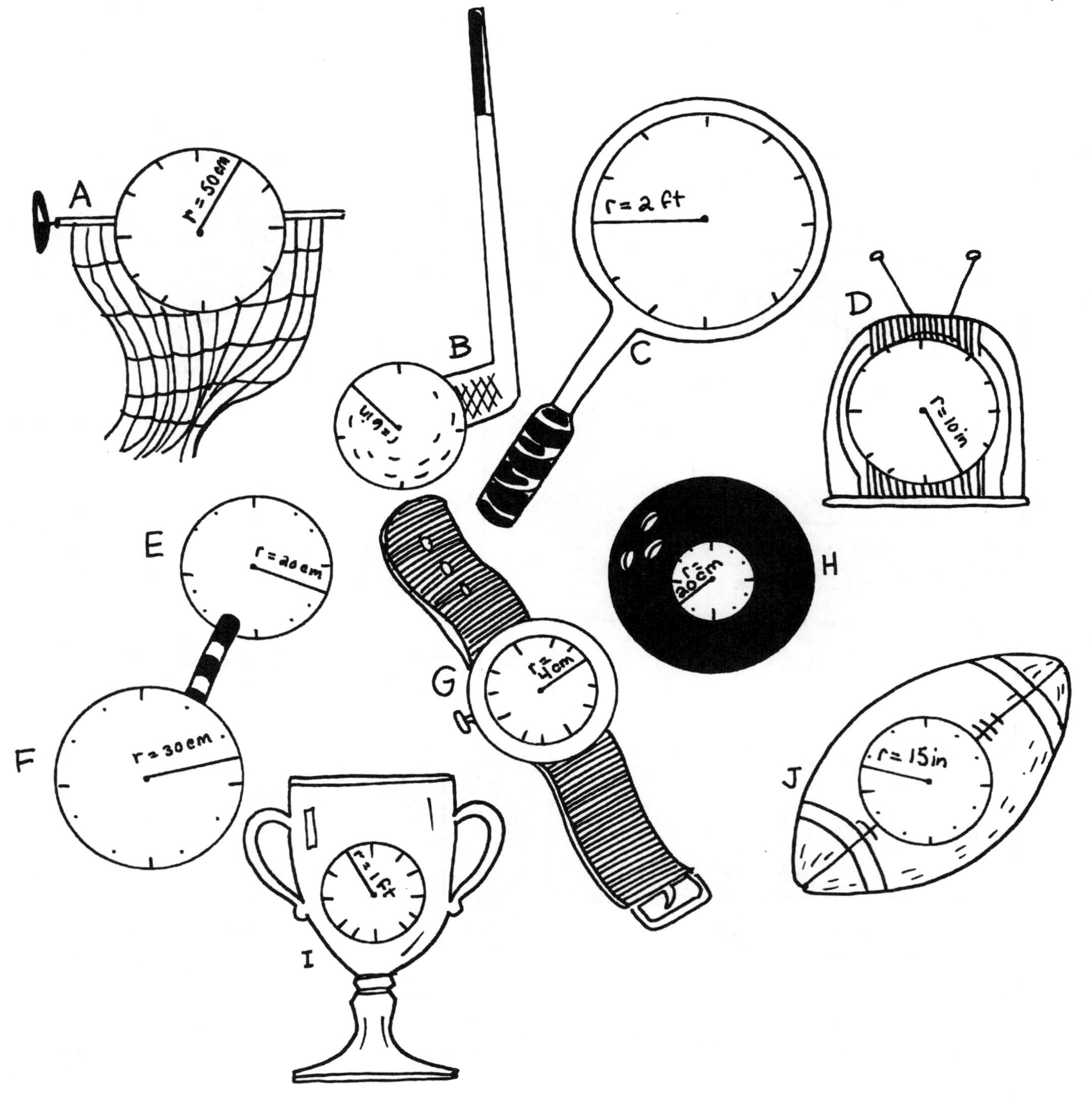

Name

SKY-HIGH MEASUREMENTS

Kites are a great way to show off your school spirit. The sky has been the limit for the Grizzly fans who got together to fly their kites before the soccer game. Find the area of each kite. Write the area on the kite or kite section.

Triangle	$A = \frac{1}{2}bh$	Rectangle	$A = l \times w$
Square	$A = s^2$	Parallelogram	$A = bh$

Name

Find Area and Perimeter

PEP RALLY MEASUREMENTS

School ended early today so students could get ready for tomorrow's pep rally. Kids are busy making posters and decorations. How much paper did they have to buy or borrow to make these posters?

To find out the sizes of the posters, calculate the perimeter and area for each one.

1. P = ______________
   A = ______________

2. P = ______________
   A = ______________

3. P = ______________
   A = ______________

4. P = ______________
   A = ______________

5. P = ______________
   A = ______________

6. P = ______________
   A = ______________

7. P = ______________
   A = ______________

8. Which has the largest area? ______________

9. Which has the longest perimeter? ______________

Name ______________________________

HUNGRY FANS

The game has gone into overtime, and the fans are extremely hungry! They are going to need a lot of snacks before the night is over.

Find the volume of each container to discover which fan got the most to eat or drink! (Volume = length x width x height) Label your answers correctly.

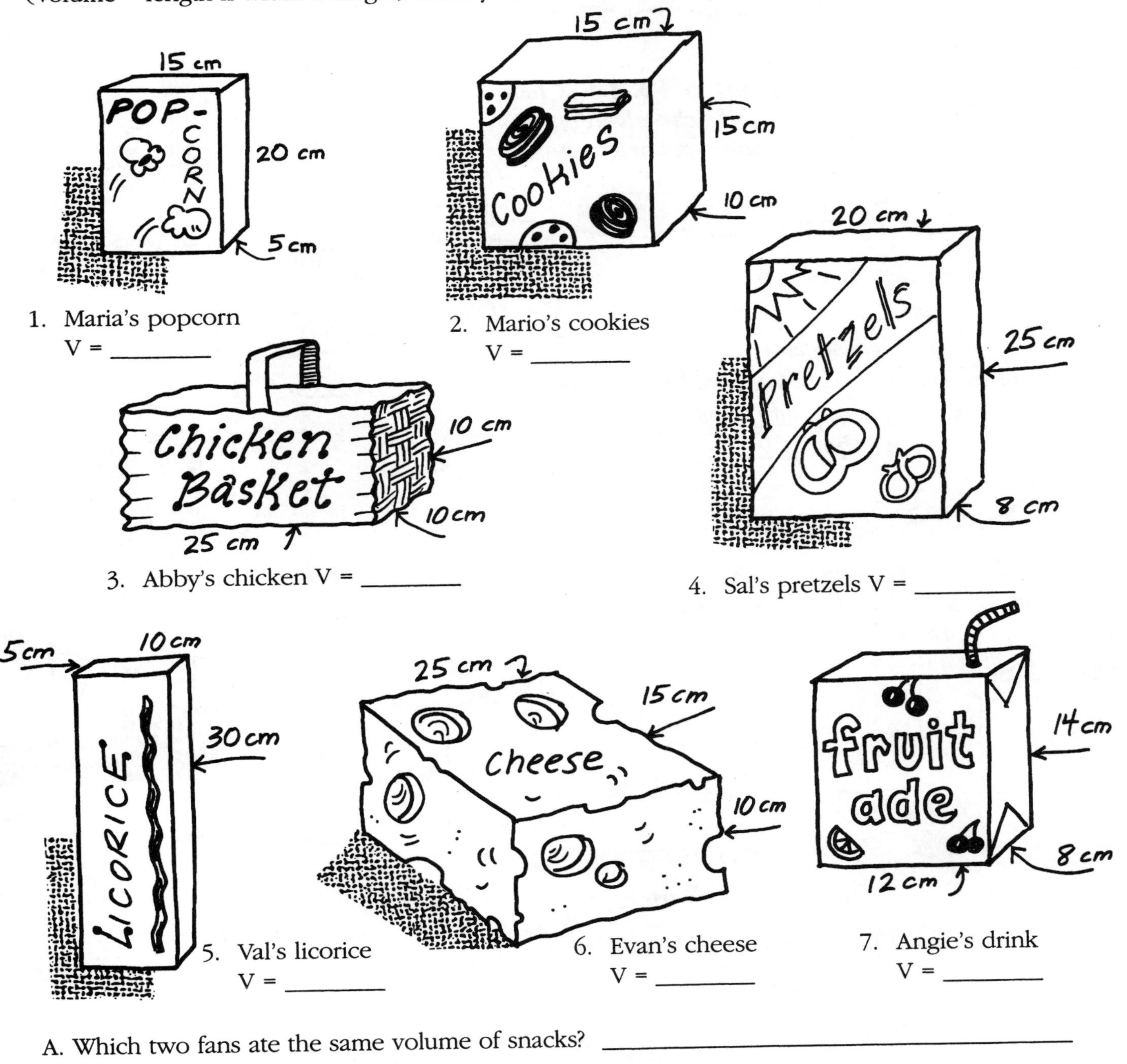

1. Maria's popcorn
   V = ________
2. Mario's cookies
   V = ________
3. Abby's chicken V = ________
4. Sal's pretzels V = ________
5. Val's licorice
   V = ________
6. Evan's cheese
   V = ________
7. Angie's drink
   V = ________

A. Which two fans ate the same volume of snacks? ____________________

B. Which fan ate or drank the greatest volume? ____________________

Name ____________________

UNIFORM CONFUSION

Al, the athletic director, is confused. He is passing out boxes of uniforms but doesn't know which one to give to each coach.

Identify each coach's box of uniforms from the description the coach is giving. Fill in the blank with the letter of the box.

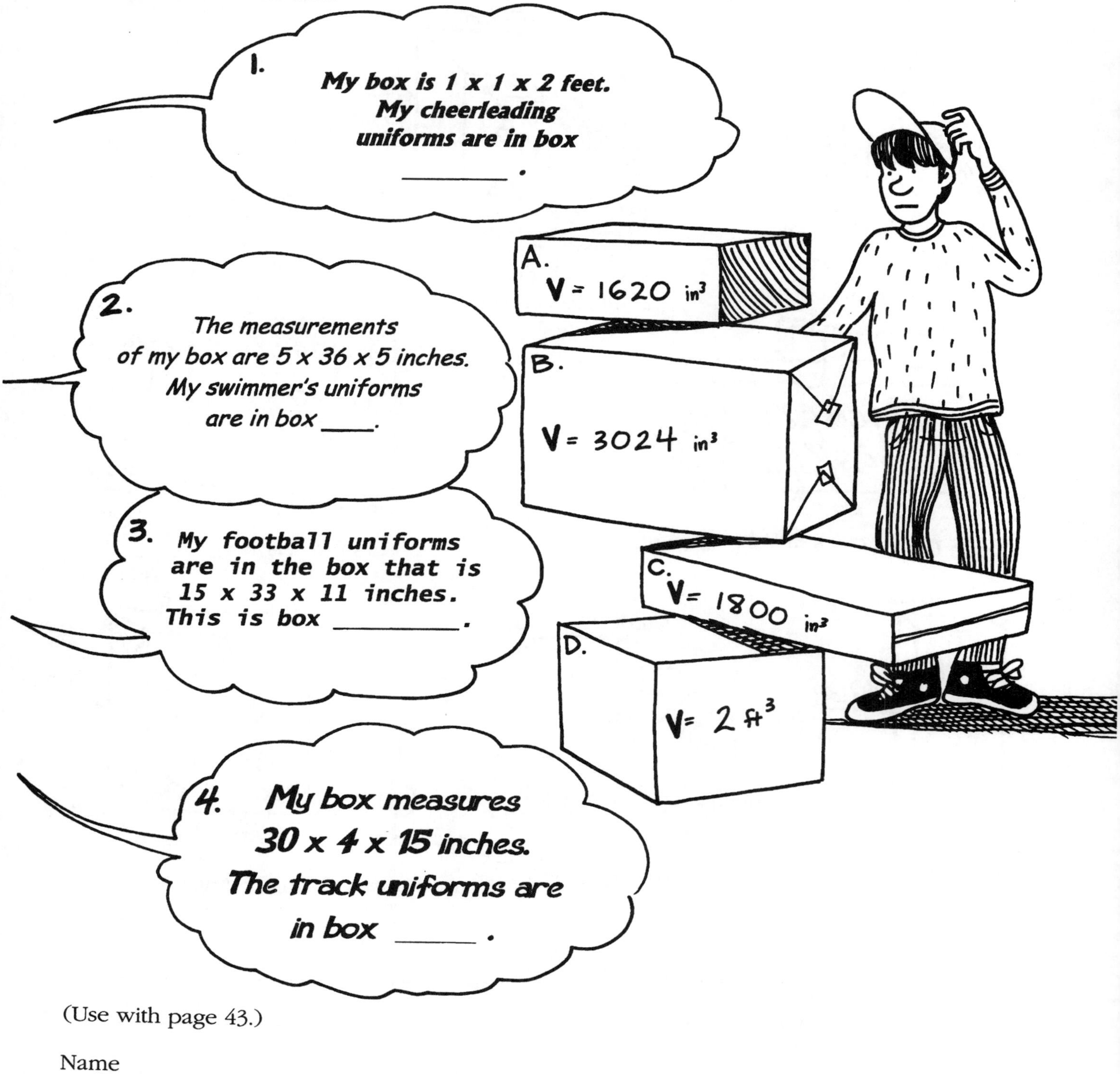

1. ***My box is 1 x 1 x 2 feet. My cheerleading uniforms are in box ________ .***

2. *The measurements of my box are 5 x 36 x 5 inches. My swimmer's uniforms are in box ____.*

3. My football uniforms are in the box that is 15 x 33 x 11 inches. This is box ________ .

4. *My box measures 30 x 4 x 15 inches. The track uniforms are in box _____ .*

A. $V = 1620\ in^3$

B. $V = 3024\ in^3$

C. $V = 1800\ in^3$

D. $V = 2\ ft^3$

(Use with page 43.)

Name

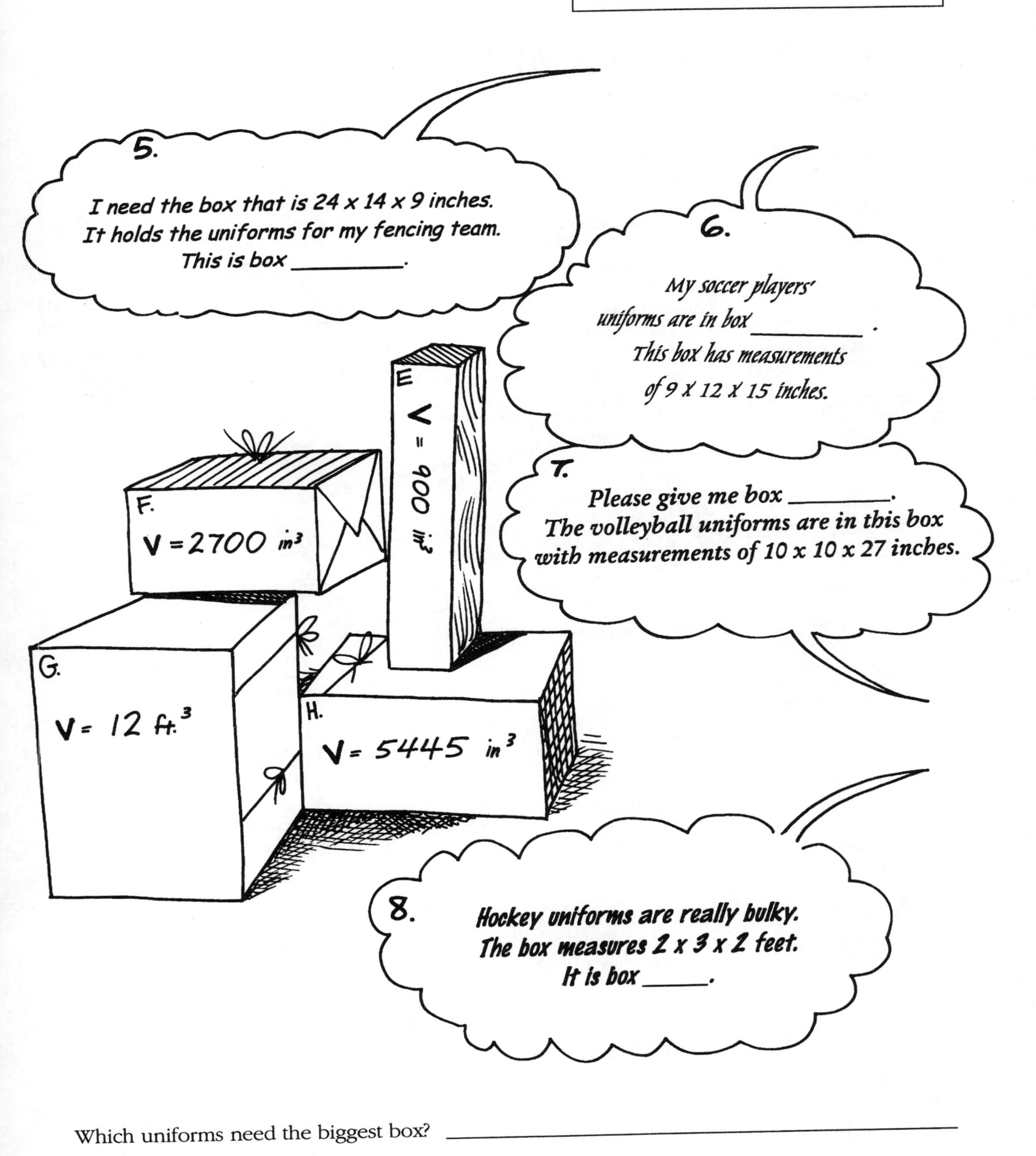

Which uniforms need the biggest box? ________________________________

(Use with page 42.)

Name ________________________________

Use Formulas to Find Perimeter, Area, and Volume

DUFFEL BAG JUMBLE

Brayden just came home from tennis practice and dumped the contents of his duffel bag on the floor of his room. Everything in his bag is a geometric plane figure or space figure. Use the right formula to find the perimeter (P), area (A), circumference (C), or volume (V) of these figures. Label each answer accurately.

FORMULAS

P = sum of all sides
C = π x d
Area of Circle = π x r^2
Area of Rectangle = l x w
Volume of Cube = S x S x S
Volume of Rectangular Prism = l x w x h
Volume of Cylinder = π x r^2 x h

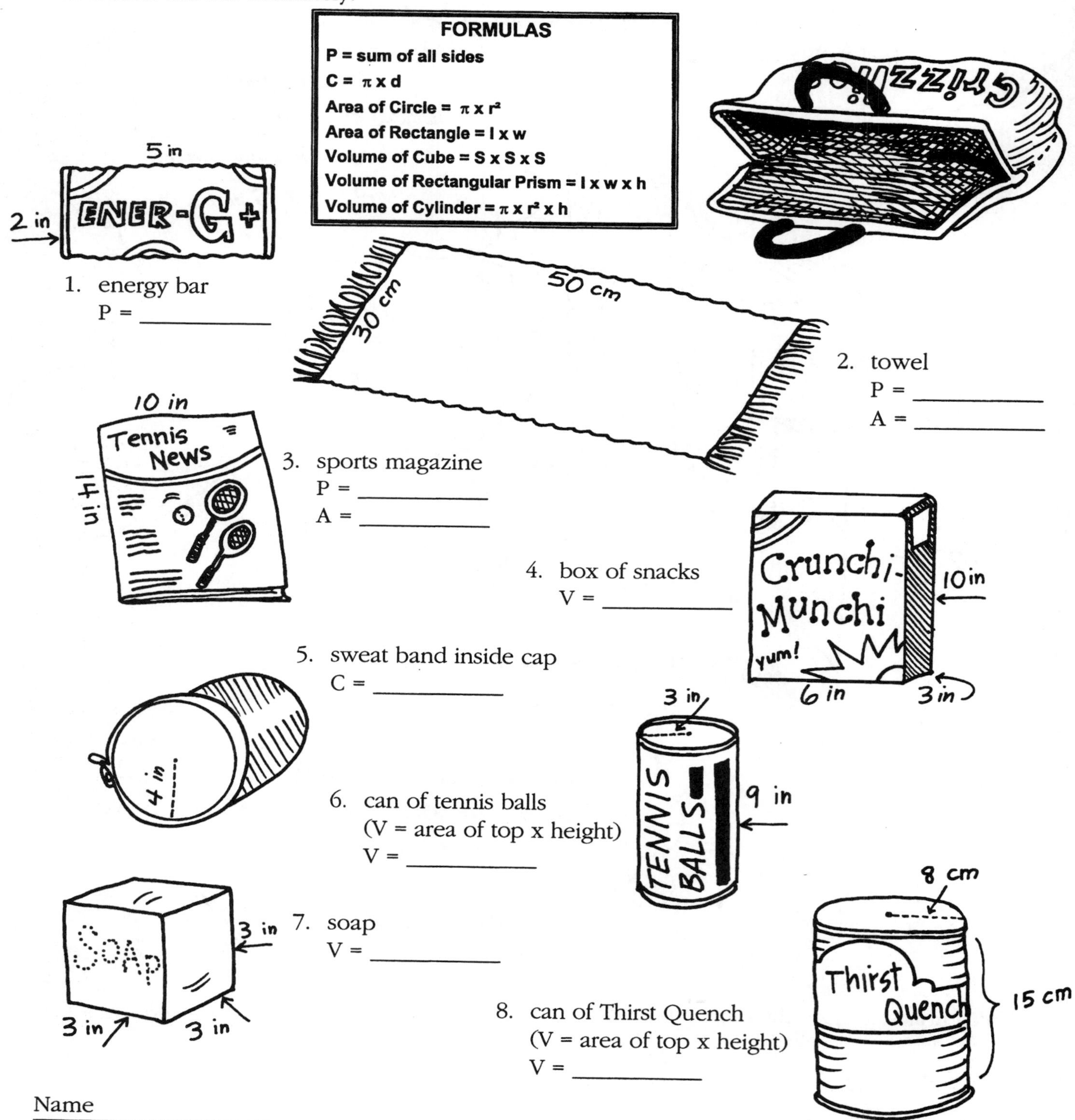

1. energy bar
   P = __________

2. towel
   P = __________
   A = __________

3. sports magazine
   P = __________
   A = __________

4. box of snacks
   V = __________

5. sweat band inside cap
   C = __________

6. can of tennis balls
   (V = area of top x height)
   V = __________

7. soap
   V = __________

8. can of Thirst Quench
   (V = area of top x height)
   V = __________

Name

LOST BALL!

On the way home from the big game in Crescent City, a basketball accidentally bounced off the bus. Trace the route that the basketball took as it rolled along around the town.

Use a metric ruler and the map scale to find out how much distance the ball traveled before it was found. Round your measurement to the nearest centimeter. Then use the map scale to convert it to meters.

Approximately how far did the ball travel?

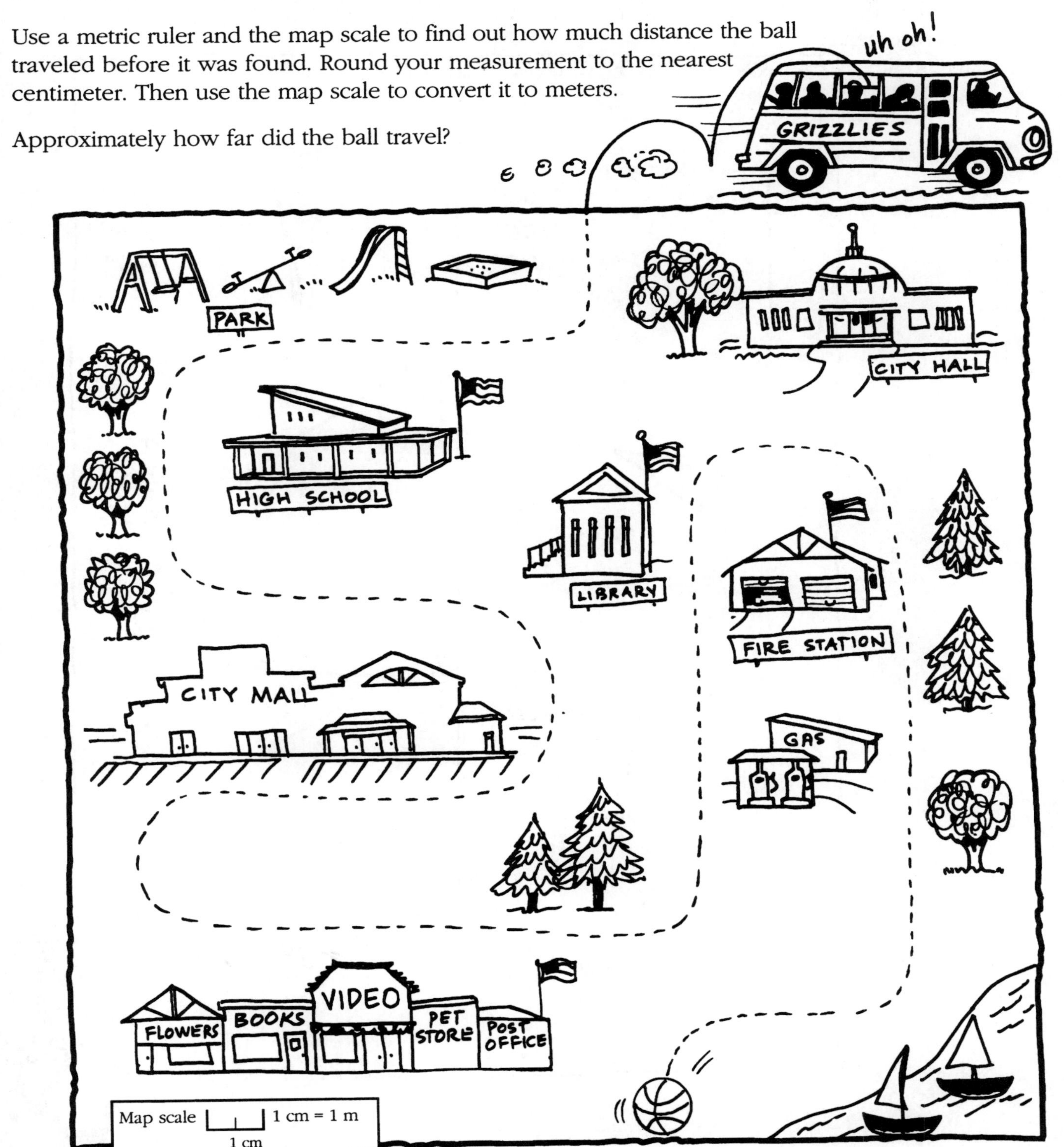

Name

ANGLES AT THE POOL

For divers, good form is very important. They must hold their bodies at exactly the right angles! Deva keeps a chart of the body positions for her different dives.

Look at the angles shown on the chart. Then take the plunge, and circle the best estimate for the measurement of each angle.

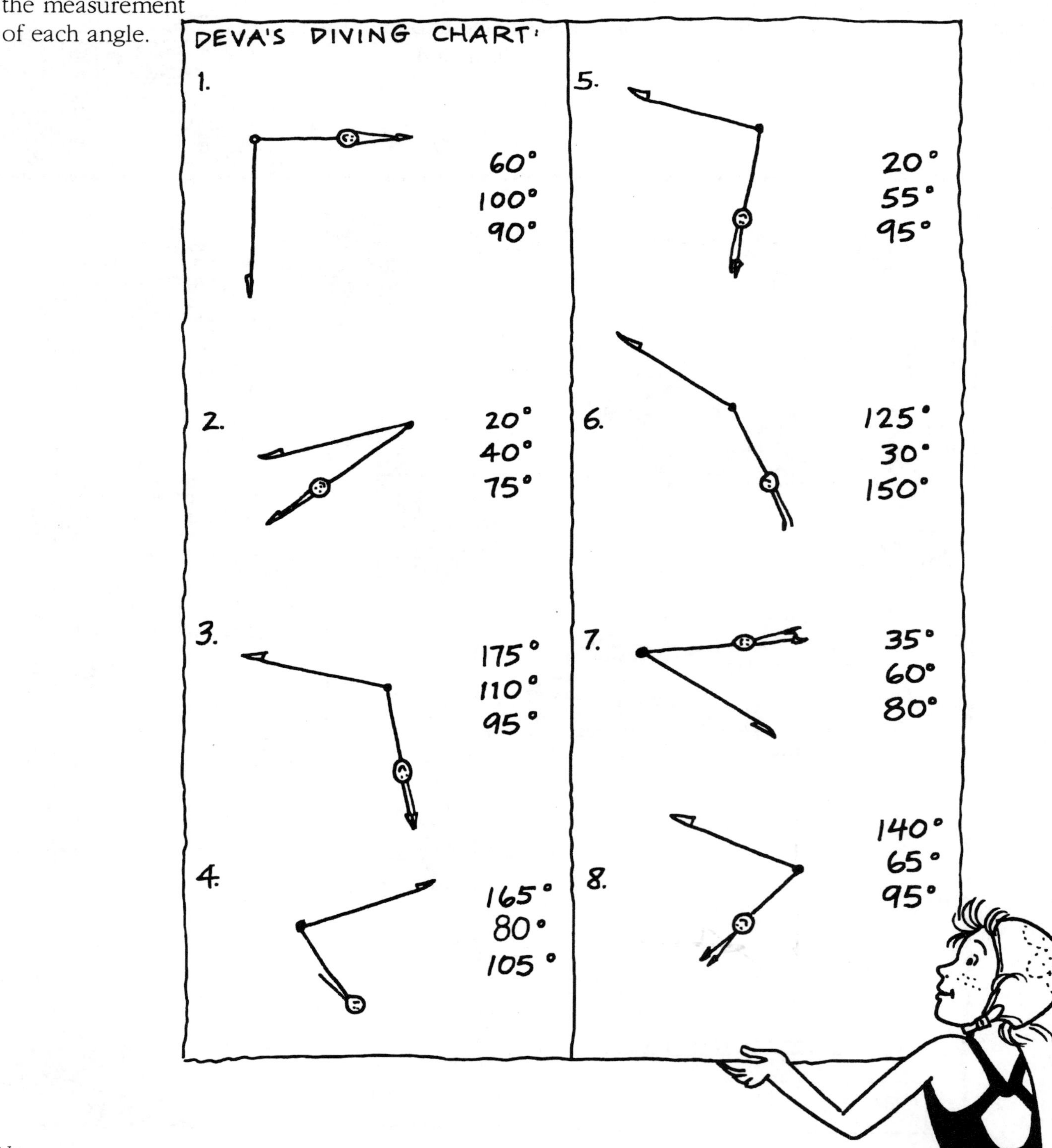

Name

ANGLES AT THE GYM

Angles show up a lot at the gym, too. Gymnasts must pay attention to good form. Notice the sharp angles that gymnasts hold as they perform their routines.

Use a protractor to measure each angle shown on the chart. Write the measurement to the nearest whole degree.

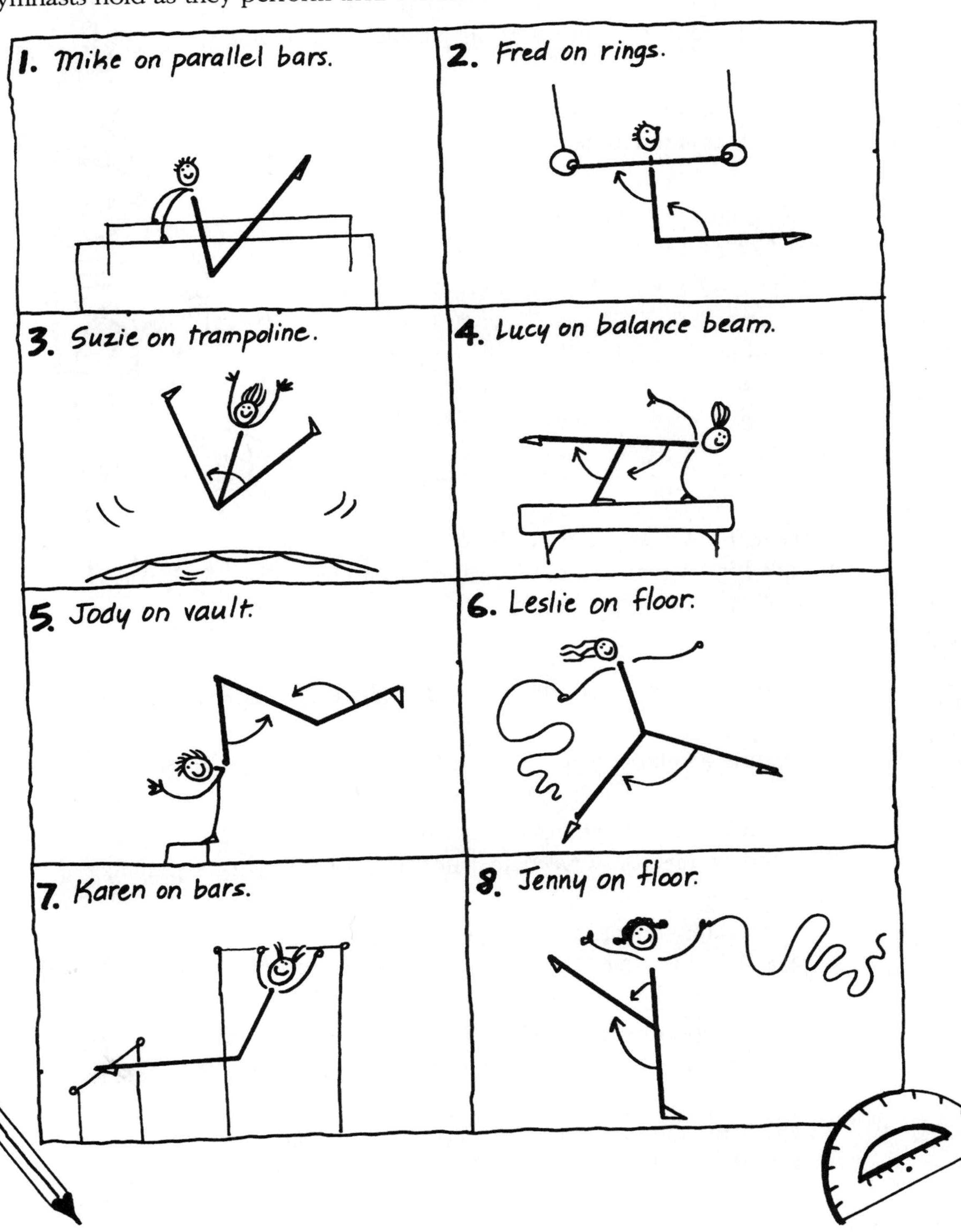

Name

JUGGLING THE SCHEDULE

Brianna is trying to juggle a very heavy schedule. She is busy with basketball, school, and other activities. Use the calendar on the next page to answer questions about her schedule.

1. What is the longest time Brianna has between basketball games in February? 8days
2. Team pictures will be taken 1 week and 1 day after her big math test. What day will that be? wednesday
3. If Brianna's history project was assigned 2 weeks and 3 days before it was due, on what day did her teacher assign it?
   Monday February 2
4. How many weeks are there between her mom's birthday and her baby-sitting job?
   3
5. How many days before the Valentine Dance did she ask Jay to be her date? 12
6. Is Brianna available to baby-sit in the evening 21 days after the team pizza party? yes
7. If soccer practice begins 2 weeks and 4 days after the tryouts, on what date will Brianna begin soccer practice? March 18th
8. Brianna plans to start studying for her math test 5 days ahead of time. On what date will she begin her studying? February 5
9. A dentist appointment is scheduled for 15 days after the Science Fair. What day of the week will that be? tuesday
10. What is the longest number of days that Brianna went without a basketball practice in February? 4

(Use with page 49)

Name ______

FEBRUARY

Sunday	Monday	Tuesday	Wednesday	Thursday	Friday	Saturday
1	2 Ask Jay to go to the Valentine Dance. • Practice – 3:00 - 4:30	3 Game! 6:00- 8:00	4 • Basketball • Practice – 3:00 - 4:30	5 MOM'S BIRTHDAY • Basketball • Practice – 3:00 - 4:30	6	7 Basketball Team's PIZZA PARTY 7:00 PM
8	9 • Basketball • Practice – 3:00 - 4:30	10 Game! 6:00 - 8:00 Math Test	11 BB • Practice 3:00 - 4:30	12 • Basketball • Practice – 3:00 - 4:30	13 Game! 6:00 - 8:00	14 Valentine Dance ! # !
15 Go Shopping	16 NO SCHOOL Presidents' Day	17 • Basketball Practice – 3: - 4:30	18 • Basketball • Practice – 3:00 - 4:30	19 History Project Due !! (uh oh) • Practice 3:- 4:30	20 Sleep over at Katie's Dance Practice	21 BIG GAME
22	23 Science Fair Basketball Practice 3: - 4:30	24 Game 6:- 8:00	25 • Basketball Practice 3:- 4:30	26 babysit for the Myers – 6:00 - 8:00 BB practice – 3:- 4:30	27 REPORT CARDS GAME 6:- 8:00	28 Soccer tryouts 9 am.

11. How long after Brianna's math test will she get a report card? 2 weeks and 3 days or 17 days

12. From the date of the Science Fair, in how many days will it be March 17? 27

13. What day of the week will it be 2 weeks and 6 days from February 27?

14. What will the date be 3 weeks and 4 days from February 16? March [illegible]

(Use with page 48.)

Name

Measure Time

THE LONGEST PRACTICES

The Ashland Middle School Marching Band practices longer than any sports team. They work all through the year to be an award-winning band.

Use your skills with time measurement to figure out just how long and hard they work! Finish the chart to show the length of each summer practice.

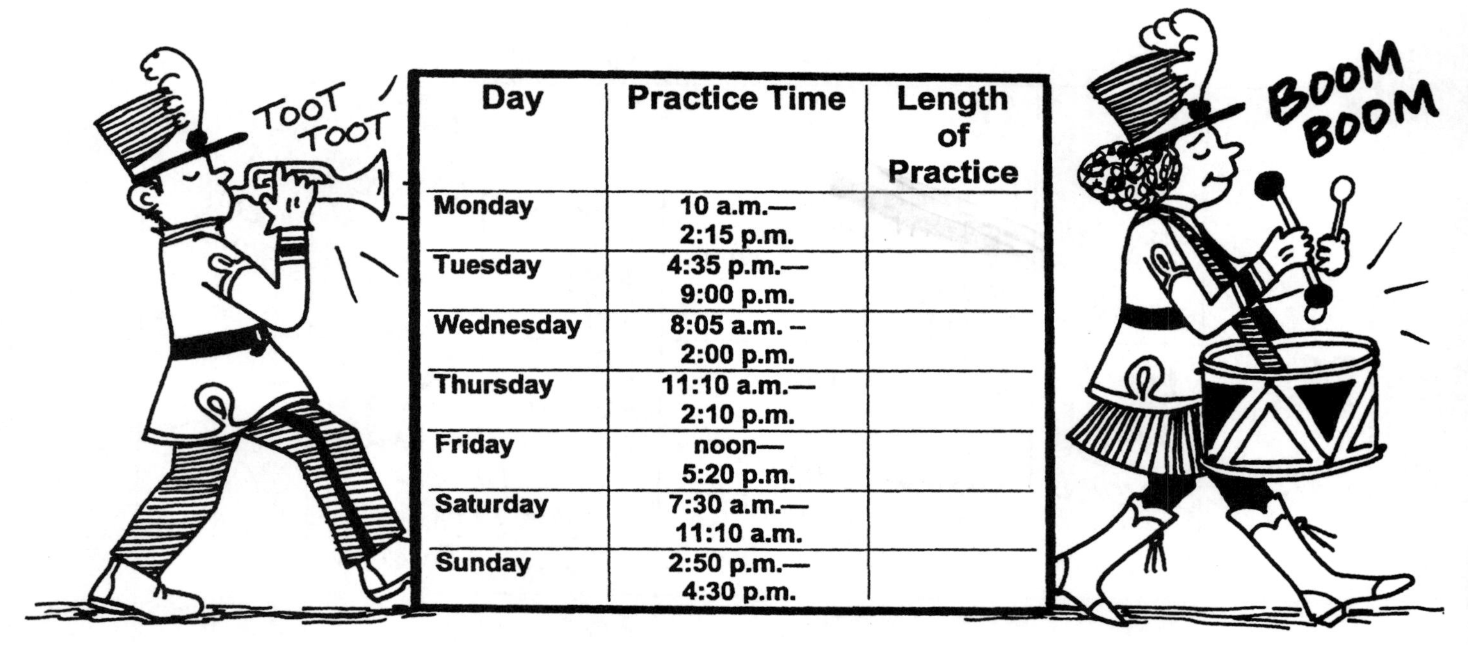

Day	Practice Time	Length of Practice
Monday	10 a.m.—2:15 p.m.	
Tuesday	4:35 p.m.—9:00 p.m.	
Wednesday	8:05 a.m. – 2:00 p.m.	
Thursday	11:10 a.m.—2:10 p.m.	
Friday	noon—5:20 p.m.	
Saturday	7:30 a.m.—11:10 a.m.	
Sunday	2:50 p.m.—4:30 p.m.	

Solve these problems.

1. During February, each practice began at 3:45 p.m. and finished at 6:10 p.m. on school days. How long was each practice? 4 hours______________
2. On Saturdays, the band warms up in the band room for 40 minutes. Then they practice on the field for 2 hours and 20 minutes. Then they go back to the band room for another 55 minutes. How long is the practice? ______________
3. When the band traveled to the latest competition, they left school on the bus at 9:30 a.m. They arrived in Astoria and practiced until 7:15 p.m. How long did they spend traveling and practicing that day? ______________
4. The trumpet section needed some extra practice. They began at 7:15 a.m. Their practice lasted 1 hour, 20 minutes. What time did it end? ______________
5. The tubas practiced after school for 2 hours and 25 minutes. They ended their practice at 4:30 p.m. What time did the practice begin? ______________

Name ______________________________

APPENDIX

Contents

GLOSSARY OF GEOMETRY AND MEASUREMENT TERMS

ALTITUDE OF A TRIANGLE — The distance between a point on the base and the vertex of the opposite angle, measured along a line which is perpendicular to the base. (The altitude is also referred to as the height of the triangle.)

ANGLE — A figure formed by two rays having a common endpoint (vertex).

An ACUTE ANGLE — measures less than 90°.

A RIGHT ANGLE — measures 90°.

An OBTUSE ANGLE — measures more than 90°and less than180°.

CONGRUENT ANGLES — are angles measuring the same.

ARC — A portion of the edge of a circle between any two points on the circle.

AREA — The measure of the region inside a closed plane figure, measured in square units.

BASE — A side of a geometric figure.

CAPACITY — The measure of the amount that a container will hold.

CHORD — A line segment having endpoints on a circle.

CIRCLE — A closed curve in which all points on the edge are equidistant from a given point in the same plane.

CIRCUMFERENCE — The distance around a circle. (Circumference = π x diameter)

COMPASS — A tool for drawing circles.

CONE — A space figure with a circular base.

CONGRUENT — Of equal size. The symbol ≅ means congruent.

CUBE — A space figure having six congruent square faces.

CURVE — A set of points connected by a line segment.

CUSTOMARY UNITS — Units of the measurement system used often in the United States (ounces, pounds, inches, feet, miles).

CYLINDER — A space figure having two congruent circular bases.

DECAGON — A 10-sided polygon.

DEGREE — 1. A unit of measure used in measuring angles. (A circle contains 360 degrees.)

2. A unit of measure used in measuring temperature.

DIAMETER — A line segment which has its endpoints on the circle and which passes through the center of the circle.

DODECAHEDRON — A polygon with 12 faces.

EDGE — A line segment formed by the intersection of two faces of a geometric space figure.

ENDPOINT — A point at the end of a line segment or ray.

EQUILATERAL — Having sides of the same length.

ESTIMATE — An approximation or rough calculation.

EVEN NUMBER — One of the set of whole numbers having 2 as a factor.

FACE — A plane region serving as a side of a space figure.

FLIP — To turn over a geometric figure. The size or shape of the figure does not change.

GEOMETRY — The study of space and figures in space.

GRAM — A standard unit for measuring weight in the metric system.

HEMISPHERE — Half a sphere.

HEPTAGON — A 7-sided polygon.

HEXAGON — A 6-sided polygon.

INTERSECTION OF LINES — The point at which two lines meet.

INTERSECTION OF PLANES — A line formed by the set of points at which two planes meet.

LINE — A set of points along a path in a plane.

LINE OF SYMMETRY — A line on which a figure can be folded so that the two parts are exactly the same.

LINE SEGMENT — Part of a line consisting of a path between two endpoints.

LINEAR MEASURE (or length) — The measure of distance between two points along a line.

LITER — A metric system unit of measurement for liquid capacity.

MEASUREMENT — The process of finding length, area, capacity, or amount of something.

METER — A metric system unit of linear measurement.

METRIC SYSTEM — A system of measurement based on the decimal system.

NONAGON — A nine-sided polygon.

OCTAGON — An eight-sided polygon.

PARALLEL LINES — Lines in the same plane which do not intersect.

PARALLELOGRAM — A quadrilateral whose opposite sides are parallel.

PENTAGON — A five sided polygon.

PERIMETER — The distance around the outside of a closed figure.

PERPENDICULAR LINES — Two lines in the same plane that intersect at right angles.

PI — The ratio of a circle's circumference to its diameter. (The symbol for pi is π, and the numerical value of pi is 3.14.)

PLANE — The set of all points on a flat surface which extends indefinitely in all directions.

PLANE FIGURE — A set of points in the same plane enclosing a region.

POINT — An exact location in space.

POLYGON — A simple closed plane figure having line segments as sides.

POLYHEDRON — A space figure formed by intersecting plane surfaces called faces.

PRISM — A space figure with two parallel congruent polygonal faces (called bases). The prism is named by the shape of its bases.

PROTRACTOR — An instrument used for measuring angles.

PYRAMID — A space figure having one polygonal base and triangular faces which have a common vertex.

QUADRILATERAL — A four-sided polygon.

RADIUS — A line segment having one endpoint in the center of the circle and the other on the circle.

RAY — A portion of a line extending from one endpoint indefinitely in one direction.

RECTANGLE — A parallelogram having four right angles.

REGION — The set of all points on a closed curve and its interior.

RHOMBUS — A parallelogram having congruent sides.

SEGMENT — Two points and all points between them.

SIMILARITY — A property of geometric figures having angles of the same size.

SIMPLE CLOSED CURVE OR FIGURE — A closed curve whose path does not intersect itself.

SLIDE — Moving a figure without turning or flipping it. The shape or size of a figure is not changed by a slide.

SPHERE — A space figure formed by a set of points lying equidistant from a center point.

SQUARE — A rectangle with all sides congruent.

SURFACE — A region lying on one plane.

SURFACE AREA — The space covered by a plane region or by the faces of a space figure.

SYMMETRIC FIGURE — A figure having two halves that are reflections of one another. A line of symmetry divides the figure into two congruent parts.

TRAPEZOID — A quadrilateral having only two parallel sides.

TRIANGLE — A three-sided polygon.

ACUTE TRIANGLE — A triangle in which all three angles are less than 90°.

EQUILATERAL TRIANGLE — A triangle having all sides and angles equal.

ISOSCELES TRIANGLE — A triangle with at least two congruent sides.

OBTUSE TRIANGLE — A triangle having one angle greater than 90°.

RIGHT TRIANGLE — A triangle having one 90° angle.

SCALENE TRIANGLE — A triangle in which no two sides are congruent.

TURN — A move in geometry which involves turning but not flipping a figure. The size or shape of the figure is not changed by a turn.

U.S. CUSTOMARY SYSTEM — A system of measurement used often in the United States.

VOLUME — The measure of capacity or space enclosed by a space figure.

TABLE OF MEASUREMENTS

METRIC SYSTEM		U. S. CUSTOMARY SYSTEM
1 centimeter (cm) = 10 millimeters (mm)	**LENGTH**	1 foot (ft) = 12 inches (in)
1 decimeter (dm) = 10 centimeters (cm)		1 yard (yd) = 36 inches (in)
1 meter (m) = 10 decimeters (dm)		1 yard (yd) = 3 feet (ft)
1 meter (m) = 100 centimeters (cm)		1 mile (mi) = 5280 feet (ft)
1 meter (m) = 1000 millimeters (mm)		1 mile (mi) = 1760 yards (yd)
1 kilometer (km) = 1000 meters (m)		
1 sq meter (m^2) = 10,000 sq centimeters (cm^2)	**AREA**	1 sq foot (ft^2) = 144 sq inches (in^2)
1 sq meter (m^2) = 100 sq decimeters (dm^2)		1 sq yard (yd^2) = 9 sq feet (ft^2)
1 sq kilometer (km^2) = 1,000,000 sq meters (m^2)		1 acre (a) = 4840 sq yards (yd^2)
		1 sq mile (mi^2) = 640 acres (a)
1 cu meter (m^3) = 1,000,000 cu centimeters (cm^3)	**VOLUME**	1 cu foot (ft^3) = 1728 cu inches (in^3)
1 cu meter (m^3) = 1000 cu decimeters (dm^3)		1 cu yard (yd^3) = 27 cu feet (cu^3)
1 cu meter (m^3) = 1 liter (L)		1 cu yard (yd^3) = 46,656 cu inches (in^3)
1 teaspoon = 5 milliliters (mL)	**CAPACITY**	1 tablespoon (T) = 3 teaspoons (t)
1 tablespoon = 12.5 milliliters (mL)		1 cup (c) = 16 tablespoons (T)
1 liter (L) = 1000 milliliters (mL)		1 cup (c) = 8 fluid ounces (fl oz)
1 liter (L) = 1,000 cu centimeters (cm^3)		1 pint (pt) = 2 cups (c)
1 kiloliter (kL) = 1000 liters (L)		1 quart (qt) = 4 cups (c)
		1 quart (qt) = 2 pints (pt)
		1 gallon (gal) = 16 cups (c)
		1 gallon (gal) = 8 pint (pt)
		1 gallon (gal) = 4 quarts (qt)
1 gram = 1000 milligrams (mg)	**WEIGHT**	1 pound (lb) = 16 ounces (oz)
1 kilogram (kg) = 1000 grams (g)		1 ton (T) = 2000 pounds (lb)
1 metric ton (t) = 1000 kilograms (kg)		
1 minute (mins) = 60 seconds (sec)	**TIME**	1 minute (mins) = 60 seconds (sec)
1 hour (hr) = 60 minutes (min)		1 hour (hr) = 60 minutes (min)
1 day = 24 hours (hr)		1 day = 24 hours (hr)
1 week = 7 days		1 week = 7 days
1 year (yr) = 365–366 days or 52 weeks		1 year (yr) = 365–366 days or 52 weeks
1 decade = 10 years (yr)		1 decade = 10 years (yr)
1 century = 100 years (yr)		1 century = 100 years (yr)
1 millennium = 1000 years (yr)		1 millennium = 1000 years (yr)

FORMULAS

PERIMETER	
Triangle	$P = a + b + c$
Rectangle	$P = 2(h + w)$
Circle (Circumference)	$P = \pi d$

AREA	
Circle	$A = \pi r^2$
Square	$A = s^2$
Triangle	$A = \frac{1}{2} bh$
Rectangle	$A = l \times w$
Trapezoid	$A = \frac{1}{2}(b1 + b2)h$

VOLUME	
Cube	$V = s^3$
Rectangular Prism	$V = B \times h$ *(B = area of base)*
Triangular Prism	$V = B \times h$ *(B = area of base)*
Pyramid	$V = \frac{1}{2} Bh$ *(B = area of base)*
Sphere	$V = \frac{4}{3}\pi r^3$
Cylinder	$V = \pi r^2 h$
Cone	$V = 1/3 \pi r^2 h$

GEOMETRY & MEASUREMENT
SKILLS TEST

(You will need a protractor and centimeter/inch ruler.)

Use this diagram for questions 1-7.

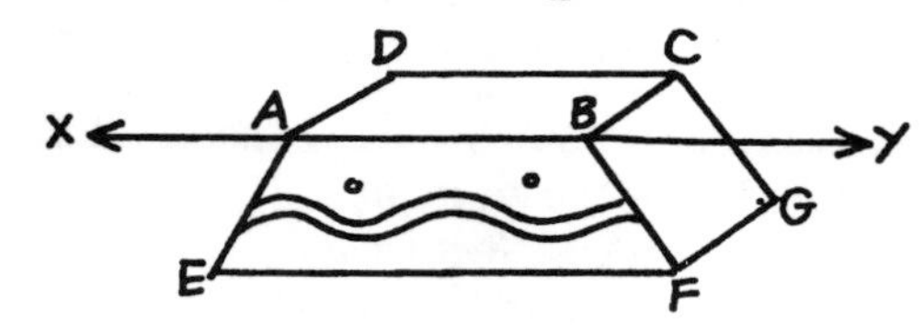

1. XY is: a line segment a ray a line a plane
2. Which ones of these are line segments?
   AB BF EA AC FY CG
3. AY is: a line segment an angle a line a ray
4. B is: a line a point a ray
5. Which pairs of line segments are parallel?
   AB & CD CD & BF EF & AB
   AD & BC BC & FG DC & FG
6. EAB is: a plane a line an angle a line segment
7. BCGF is: a line segment an angle a plane a line

Use these angles for questions 8–13.

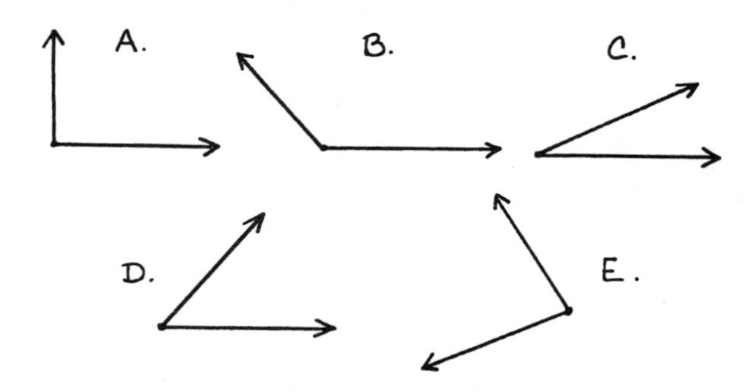

8. Which angles are right angles?

   ____

9. Which angles are obtuse angles?

   ____

10. Which angles are acute angles?

    ____

11. Angle C is about:
    30° 130° 95° 50°
12. Write the measurement of angle B.

    ____

13. Which angle is about 80°?

    ____

Use the diagram below for questions 14-19.

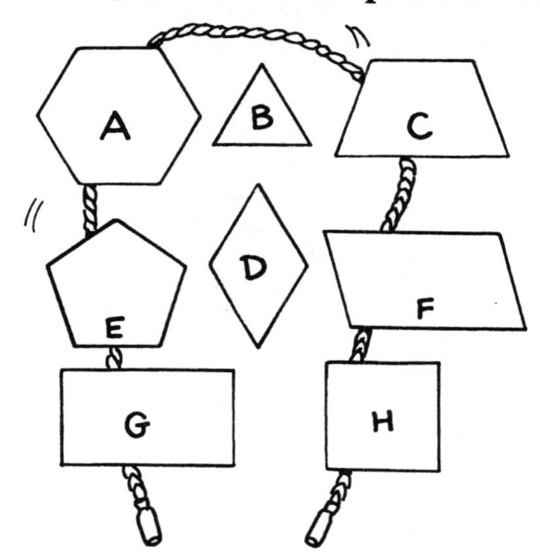

_______ 14. How many figures are rhombuses?

_______ 15. Which figure is a hexagon?

_______ 16. Which figures are parallelograms but not rectangles?

_______ 17. Which figure is a pentagon?

_______ 18. Which figure is a trapezoid?

_______ 19. Which figures are rectangles?

Name ______________________________

Use the diagram below for questions 20–23.

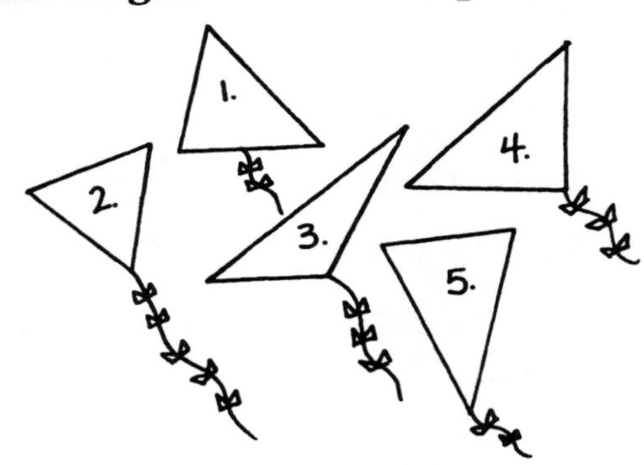

_______ 20. Which ones are scalene triangles?

_______ 21. Which are isosceles triangles?

_______ 22. Which are right triangles?

_______ 23. Which are equilateral triangles?

Use the diagram below for questions 24–27.

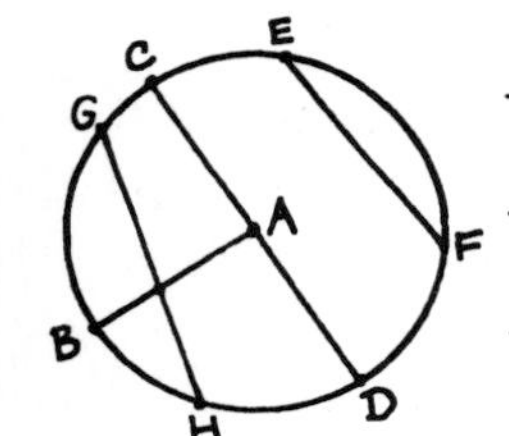

_______ 24. Name the diameter.

_______ 25. Name three chords.

_______ 26. Name three radii.

_______ 27. Is EA an arc?

Use the signs below for questions 28–32.

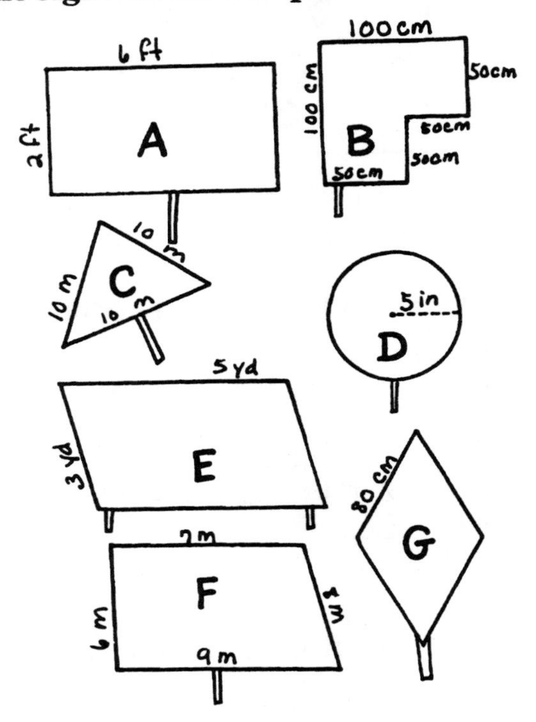

_______ 28. Find the perimeter of E.

_______ 29. Find the perimeter of G.

_______ 30. Is the perimeter of F > the perimeter of C?

_______ 31. Find the circumference of D.

_______ 32. Which has a perimeter of 400cm?

Use the pictures below for questions 33–35.

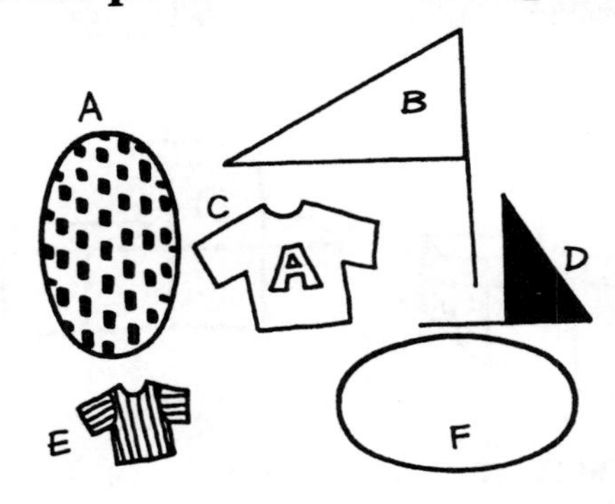

Answer true or false.

_______ 33. F is congruent to A.

_______ 34. D is congruent to B.

_______ 35. E is similar to C.

Use these diagrams for questions 36–38.

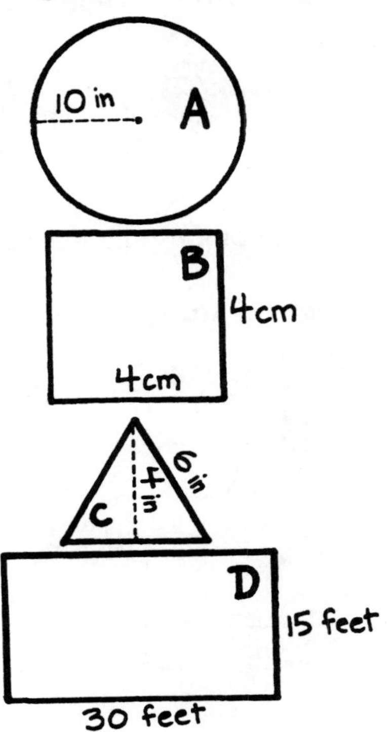

Name

_______ 36. Find the area of figure A.

_______ 37. Which figure has an area of 12 in≤?

_______ 38. Find the area of figure D.

Use this figure for questions 39–40.

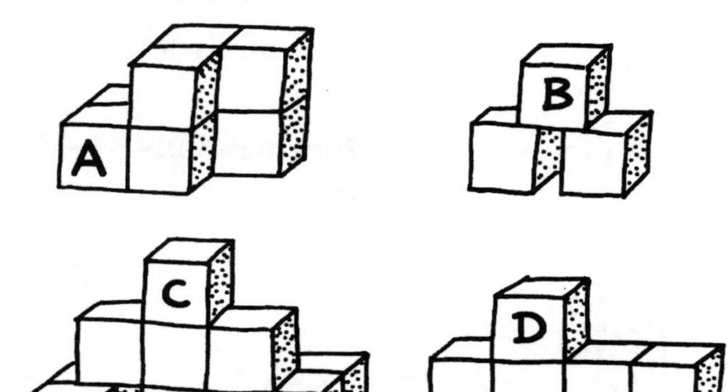

_______ 39. Which figure has a volume of 8 cubic units?

_______ 40. Which has a greater volume: D or A?

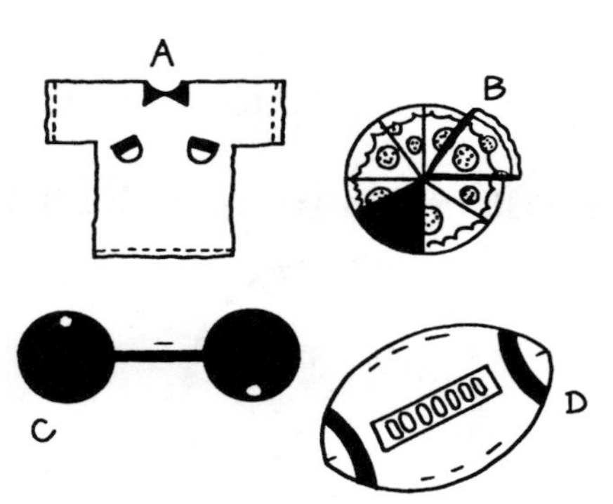

_______ 41. Which of these figures are symmetrical?

Use these figures for questions 42–47.

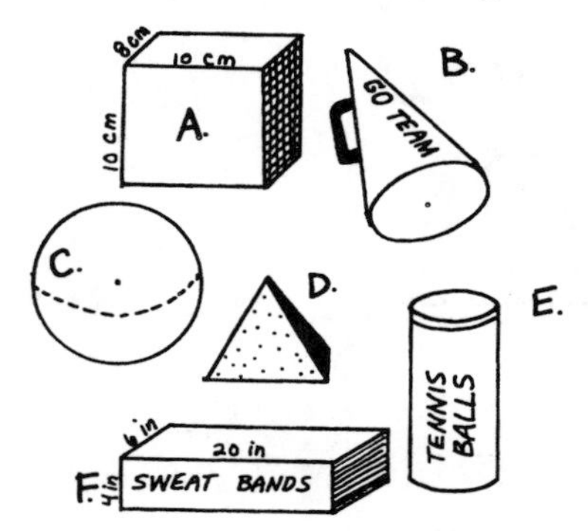

_______ 42. Which figure is a cylinder?

_______ 43. Which figure is a cone?

_______ 44. Which figure is a sphere?

_______ 45. Which figure is a pyramid?

_______ 46. Find the volume of the cube.

_______ 47. Find the volume of the rectangular prism.

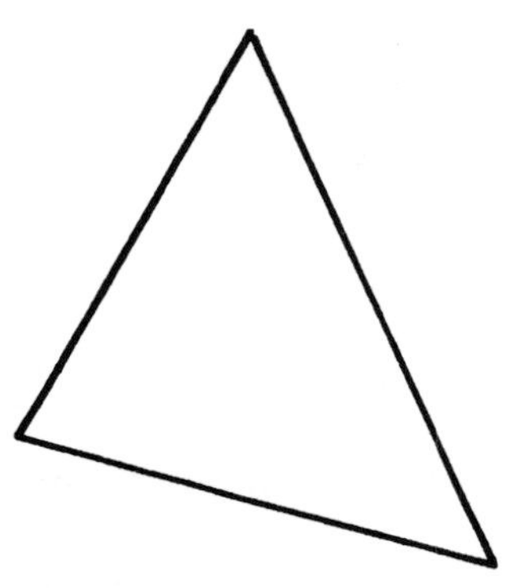

_______ 48. Measure to find the perimeter of this figure in centimeters. Round to the nearest whole centimeter.

_______ 49. Measure with centimeters to find the area of this figure.

Use these figures to answer questions 50–51.

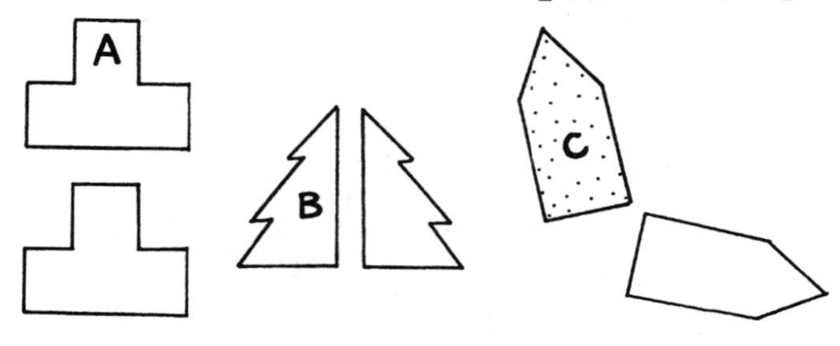

_______ 50. Which figure has been flipped?

_______ 51. Does A show a slide, flip, or turn?

Name

Circle the correct answer for questions 52–60.

52. A triangle with only two equal sides is:
    a. an equilateral triangle
    b. an isosceles triangle
    c. a scalene triangle

53. Lines that do not touch each other are:
    a. intersecting lines
    b. perpendicular lines
    c. parallel lines

54. A four-sided polygon is:
    a. a pentagon
    b. an octagon
    c. a quadrilateral
    d. a hexagon

55. A triangular pyramid has:
    a. 3 faces, 2 vertices, and 3 edges
    b. 4 faces, 1 vertex, and 3 edges
    c. 4 faces, 1 vertex, and 6 edges
    d. 3 faces, 3 vertices, and 6 edges

56. Which would be the best unit to measure the amount of water in a swimming pool?
    a. cups
    b. cubic inches
    c. liters
    d. tons

57. Which would be the best unit to measure the volume of a refrigerator?
    a. grams
    b. ounces
    c. square millimeters
    d. cubic feet

58. Which would be the best unit to measure the length of trip?
    a. square inches
    b. liters
    c. cubic feet
    d. kilometers

59. Which tool would be best for measuring the length of a running shoe?
    a. a scale
    b. a ruler
    c. a protractor
    d. a thermometer

60. Which tool would be best for measuring the time from now until the next New Year's Eve?
    a. a stopwatch
    b. a yardstick
    c. a calendar
    d. a thermometer
    e. a clock

Write T (*true*) or F (*false*) for statements 61–65.

______ 61. A rhombus is always a square.

______ 62. All rectangles have 4 right angles.

______ 63. An octagon has 6 equal sides.

______ 64. A square is a rectangle.

______ 65. A rectangle is a quadrilateral.

Fill in each blank with the correct amount. (66–75).

66. 2 minutes = ________ seconds.
67. 1 meter = ________ centimeters.
68. 1000 meters = ________ kilometer(s).
69. 1 liter = ________ milliliters.
70. 1 pound = ________ ounces.
71. 2 quarts = ________ cups.
72. 2000 grams = ________ kilograms.
73. ________ yards = 360 inches.
74. 10 hours = ________ minutes.
75. 6 weeks = ________ days.

For each blank, write >, <, or =.

76. 10 qt ________ 2 gal.
77. 1000 mg ________ 1 g.
78. 1 T ________ 3000 pounds
79. 2 months ________ 12 weeks.
80. 10,000 m ________ 1 km.

Name ______________________________

ANSWER KEY

Skills Test

1. a line
2. AB, BF, EA, CG
3. a ray
4. a point
5. AB & CD, AD & BC, BC, EF, AB & FG
6. an angle
7. a plane
8. A
9. B
10. C, D, E
11. 50°
12. 130°
13. E
14. 2
15. A
16. D, F
17. E
18. C
19. G, H
20. 1, 3
21. 5
22. 4
23. 2, 4
24. CD
25. CD, EF, GH
26. AB, AC, AD
27. no
28. 16 yds
29. 320 cm
30. no
31. 31.4 in
32. B
33. true
34. false
35. true
36. 314 in^2
37. C
38. 450 ft^2
39. A or C
40. A
41. A, C, D
42. E
43. B
44. C
45. D
46. 800 cm^3
47. 480 in^3
48. 13 cm
49. 9cm^2
50. B
51. slide
52. b
53. c
54. c
55. c
56. c
57. d
58. d
59. b
60. c
61. false
62. true
63. false
64. true
65. true
66. 120
67. 100
68. 1
69. 1000
70. 16
71. 8
72. 2
73. 10
74. 600
75. 42
76. $>$
77. $=$
78. $<$
79. $<$
80. $>$

Skills Exercises

page 10

1. D
2. C
3. A
4. F
5. B
6. E
7. G

8–10. Check to see that students have accurately drawn the three kinds of line pairs.

page 11

1. Plane ABCD
2. Ray AB
3. Line Segment AB
4. Point B
5. Line AB
6. Angle ABC
7. Angle CBD
8. Angle ACB
9. Line Segment BC
10. Angle DCE

page 12

1. 19 line segments (May be named with letters in reverse)
   PR PQ QR RO OU
   RU NU RN PS PM
   SM SQ MQ QT TR
   SU ST TU NO
2. 32 angles (students may not find all of these). Angles may be named with letters in reverse.

–RTQ	–SPQ
–TRU	–RUT
–RTU	–QTS
–PQS (or PQM)	–RQT
–QST (or MST)	–QTU
–PSQ (or PSM)	–QRT
–SQT (or MQT)	–PST
–RQS (or RQM)	–QRU (or QRN)
–RTS	–PQT
–PMQ	–PQM
–MPS	–PMS
–NRO	–RON
–ONR	–TRO
–QRO	–ROU
–NOU	–OUN
–UNO	–TUO

3. Check to see that student has written initials in the vertex of angle RNO.

page 13

There are several possible answers for this page, particularly for line segments, points, and angles. Check to see that the student has traced or colored at least three of each in the correct color.

page 14

1. F, K, Q
2. B, H, J, L
3. A, G, O, P, E, C

page 15

There are many possible choices. Check to see that the student has traced the correct number of each type of angle in the correct color.

page 16

There are several correct choices for pairs of congruent angles. Check to see that student has traced 6 or more pairs that seem congruent.

1. no
2. no

page 17

Answers will vary because different groupings of shapes may create different kinds of triangles. Check student designs to see that triangles are colored correctly.

page 18

A. octagon
B. hexagon
C. pentagon
D. trapezoid
E. square
F. rectangle (or square)
G. parallelogram (or square or rectangle)
H. quadrilateral
I. obtuse triangle
J. equilateral triangle
K. right triangle
L. scalene triangle
M. triangle
N. polygon
O. isosceles triangle
P. rhombus (or square)

page 19

Check to see that student has colored design accurately.

1. arrow
2. archery

page 20

1. Rae Rectangle
2. Tru Trapezoid
3. Helen Hexagon
4. Tish Triangle
5. Dee Decagon
6. Suki Square
7. Olive Octagon
8. Pat Pentagon
9. Rosa Rhombus
10. Pam Parallelogram

page 21

Top

1. rectangle
2. trapezoid
3. square
4. parallelogram
5. square or rhombus
6. rhombus

Bottom

1. T
2. T
3. T
4. F
5. F
6. T
7. T
8. T
9. F
10. F

page 22

Note: In labels for line segments and arcs, the letters may be written as shown below or they may be reversed in each case (for example, a line segment may be labeled TQ or QT).

1. N
2. TQ, MK
3. NH, NT, NQ, NK, NM
4. Any two of these: RS, MK, TQ, JQ
5. Any four of these:
   HT, HJ, HK, HQ,
   HS, HR, HM

   TJ, TK, TQ, TS,
   TR, TM, TH

   JK, JQ, JS, JR,
   JM, JH, JT

   KQ, KS, KR, KM,
   KH, KT, KJ

   QS, QR, QM, QH,
   QT, QJ, QK

   SR, SM, SH, ST,
   SJ, SK, SQ

   RM, RH, RT, RJ,
   RK, RQ, RS

   MH, MT, MJ, MK,
   MQ, MS, MR

Bottom: Check student circle to see that all 9 elements have been included and properly labeled.

page 23

Top question: neither

1. C
2. C
3. S
4. S
5. C
6. C
7. S
8. S
9. C

page 24

The symmetrical figures are: A, B, D, F, G, and H. These should be colored. Also, check to see that student has drawn a correct line of symmetry in each of these figures. At the bottom of the page, check to see that student has completed each figure to look symmetrical.

page 25

The pairs of items should be labeled:

ping pong paddles:	T
Pennants:	S and F
gloves:	F
black shoes:	F
football helmets:	F
skis:	F
ping pong balls:	S
footballs:	S
boxing gloves:	S
tennis racquets:	T or F or both
ballet slippers:	T
oars:	F

Figures at bottom: check to see that student has followed instructions accurately in drawing new figures.

page 26

Spheres (blue):
baseball, golf ball, ping pong ball, bowling ball, balls at ends of barbell

Cubes (red):
Step-Up box, energy bar box

Rectangular Prisms (yellow):
shoe box, stop sign, balance beam

Cylinders (purple):
chalk, tennis ball can, bar on barbells

Answer Key

Cones (green):
liniment container, Grizzlies megaphone

Pyramid (orange):
2000# weight

Triangular Prism (brown):
tent

See diagram below for correct red path:

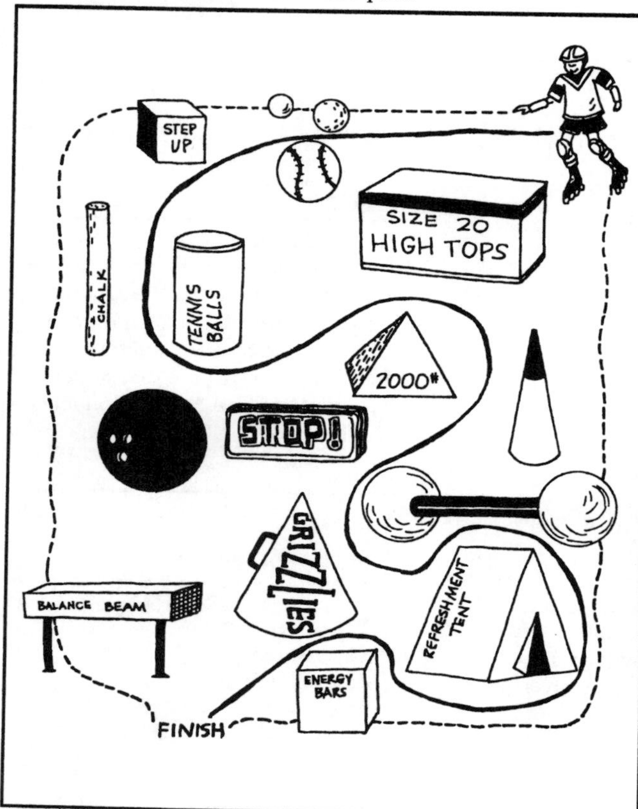

page 27

1. cube
2. triangular pyramid
3. square pyramid
4. hexagonal pyramid
5. cone
6. triangular prism
7. cylinder

page 28

1. E
2. J
3. D
4. B
5. 12
6. A, E, F, G, I, J
7. B, C, D, H, K
8. yes
9. yes
10. C
11. 1
12. G and I

page 29

Students should have the following words circled in the puzzle.

1. years
2. miles
3. minutes
4. ounces
5. feet
6. quart
7. pounds
8. hours
9. tablespoons
10. yards
11. degrees
12. seconds
13. pints
14. cups
15. inches
16. ton
17. teaspoons
18. gallons

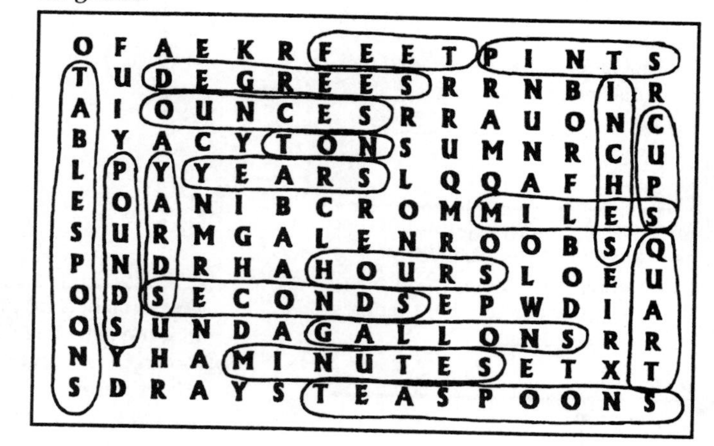

page 30

1. =
2. =
3. >
4. =
5. =
6. <
7. =
8. >
9. =
10. =
11. >
12. >
13. >
14. >
15. <
16. <
17. >
18. <
19. =
20. =

page 31

Measurements will vary. Check student answers to make sure they are reasonable and that they are written and labeled correctly with U.S. Customary units.

page 32

Circle these item numbers as correct: 2, 3, 4, 6, 7, 10, 12, 13, 16
Corrected answers are:

1. liters (or liters, milliliters, deciliters, and kiloliters)
5. 2 kilometers
8. 1000
9. 1000
11. 2000
14. 1000
15. 10,000

Answer Key

page 33

Paths students draw will vary in length. Check student's path to see that it is measured correctly, and has been correctly converted into meters, according to the scale.

page 34

1. gallon container
2. yardstick or measuring tape
3. protractor
4. thermometer
5. yardstick or measuring tape
6. scale
7. clock
8. gallon container or cup
9. scale
10. ruler or measuring tape
11. gallon container or cup
12. ruler, yardstick, or measuring tape
13. yardstick, ruler, or measuring tape
14. clock
15. scale

page 35

1. 62.8 in
2. 9.42 in
3. 4.71 in
4. 28.26 in
5. 4.71 in
6. 7.85 in
7. 6.28 in
8. 9.42 ft
9. 25.12in
10. 12.56 in
11. 18.84 cm
12. 31.4 in
13. 37.68 ft

pages 36–37

archery area .... P = 154 m
boxing area .... P = 80 ft
tennis court .... P = 228 ft
softball diamond .... P = 260 ft
track and field area .... P = 600 m
swimming pool .... P = 200 m
track .... P = 160 m
soccer field .... P = 380 yds
volleyball court .... P = 54 m

page 38

A. 7850 cm^2
B. 113.04 in^2
C. 12.56 ft^2
D. 314 in^2
E. 1256 cm^2
F. 2826 cm^2
G. 50.24 cm^2
H. 1256 cm^2
I. 3.14 ft^2
J. 706.5 in^2

page 39

1. 400 in^2
2. 12 ft^2
3. 14 m^2
4. 3600 in^2
5. 50 ft^2
6. 100,000 cm^2
7. 30 m^2
8. 24 yds^2

page 40

1. P =58 in
   A =125 in^2
2. P =10 ft
   A = 4 ft^2
3. P =136 in
   A =987 in^2
4. P =96 in
   A =576 in^2
5. P =6.5 ft
   A =1.5 ft^2
6. P = 120 in
   A = 900 in^2
7. P = 60 in
   A = 221 in^2
8. 3
9. 3

page 41

1. 1500 cm^3
2. 2250 cm^3
3. 2500 cm^3
4. 4000 cm^3
5. 1500 cm^3
6. 3750 cm^3
7. 1344 cm^3

A. Maria and Val
B. B.Sal

pages 42-43

1. D
2. E
3. H
4. C
5. B
6. A
7. F
8. G

The hockey uniforms need the largest box.

page 44

1. P = 14 in
2. P = 160 cm
   A = 1500 cm^2
3. P = 48 in
   A = 140 in^2
4. V = 180 in^3
5. C = 25.12 in
6. V = 254.34 in^3
7. V = 27 in^3
8. V = 3014.4 cm^3

page 45

Approximately 650 meters (allow answers that are close to this measurement).

page 46

1. 90°
2. 20°
3. 110°
4. 75°
5. 95°
6. 150°
7. 35°
8. 65°

page 47

Student measurements may vary slightly. Give credit for answers close to these.

1. 55°
2. both angles are 90°
3. 80°
4. angle to left is 65°; angle on right is 115°
5. angle below body is 60°; angle above legs is 130°
6. 110°
7. 120°
8. angle below leg is 130°; angle above leg is 50°

pages 48–49

1. 8 days
2. Wednesday
3. Monday, February 2
4. 3
5. 12
6. yes
7. March 18
8. February 5
9. Tuesday
10. 4
11. 2 weeks and 3 days OR 17 days
12. 22
13. Thursday
14. March 13

page 50

CHART:
Monday: 4 hours, 15 minutes
Tuesday: 4 hours, 25 minutes
Wednesday: 5 hours, 55 minutes
Thursday: 3 hours
Friday: 5 hours, 20 minutes
Saturday: 3 hours. 40 minutes
Sunday: 1 hour, 40 minutes

1. 2 hours, 25 minutes
2. 3 hours, 55 minutes
3. 9 hours, 45 minutes
4. 8:35 a.m.
5. 2:05 p.m.